Sustainable Civil Infrastructures

About this Series

Sustainable Infrastructure impacts our well-being and day-to-day lives. The infrastructures we are building today will shape our lives tomorrow. The complex and diverse nature of the impacts due to weather extremes on transportation and civil infrastructures can be seen in our roadways, bridges, and buildings. Extreme summer temperatures, droughts, flash floods, and rising numbers of freeze-thaw cycles pose challenges for civil infrastructure and can endanger public safety. We constantly hear how civil infrastructures need constant attention, preservation, and upgrading. Such improvements and developments would obviously benefit from our desired book series that provide sustainable engineering materials and designs. The economic impact is huge and much research has been conducted worldwide. The future holds many opportunities, not only for researchers in a given country, but also for the worldwide field engineers who apply and implement these technologies. We believe that no approach can succeed if it does not unite the efforts of various engineering disciplines from all over the world under one umbrella to offer a beacon of modern solutions to the global infrastructure. Experts from the various engineering disciplines around the globe will participate in this series, including: Geotechnical, Geological, Geoscience, Petroleum, Structural, Transportation, Bridge, Infrastructure, Energy, Architectural, Chemical and Materials, and other related Engineering disciplines.

More information about this series at http://www.springer.com/series/15140

Sherif Agaiby · Piergiorgio Grasso
Editors

Engineering Challenges for Sustainable Underground Use

Proceedings of the 1st GeoMEast International Congress and Exhibition, Egypt 2017 on Sustainable Civil Infrastructures

Springer

2017 GEOMEAST

SSIGE
SOIL-STRUCTURE INTERACTION GROUP IN EGYPT

Editors
Sherif Agaiby
Dar Al-Handasah (Shair and Partners)
Cairo
Egypt

Piergiorgio Grasso
Geodata Engineering
Turin
Italy

ISSN 2366-3405 ISSN 2366-3413 (electronic)
Sustainable Civil Infrastructures
ISBN 978-3-319-61635-3 ISBN 978-3-319-61636-0 (eBook)
DOI 10.1007/978-3-319-61636-0

Library of Congress Control Number: 2017946466

Printed on acid-free paper

This Springer imprint is published by Springer Nature
The registered company is Springer International Publishing AG
The registered company address is: Gewerbestrasse 11, 6330 Cham, Switzerland

Preface

Toward building sustainable and longer civil infrastructures, the engineering community around the globe continues undertaking research and development to improve existing design, modeling, and analytical capability. Such initiatives are also the core mission of the Soil-Structure Interaction Group in Egypt (SSIGE) to contribute to the ongoing research toward sustainable infrastructure. This conference series "GeoMEast International Congress and Exhibition" is one of these initiatives.

Ancient peoples built their structures to withstand the test of time. If we think in the same way, our current projects will be a heritage for future generations. In this context, an urgent need has quickly motivated the SSIGE and its friends around the globe to start a new congress series that can bring together researchers and practitioners to pursue "Sustainable Civil Infrastructures." The GeoMEast 2017 is a unique forum in the Middle East and Africa that transfers from the innovation in research into the practical wisdom to serve directly the practitioners of the industry.

More than eight hundred abstracts were received for the first edition of this conference series "GeoMEast 2017" in response to the Call for Papers. The abstracts were reviewed by the Organizing and Scientific Committees. All papers were reviewed following the same procedure and at the same technical standards of practice of the TRB, ASCE, ICE, ISSMGE, IGS, IAEG, DFI, ISAP, ISCP, ITA, ISHMII, PDCA, IUGS, ICC, and other professional organizations who have supported the technical program of the GeoMEast 2017. All papers received a minimum of two full reviews coordinated by various tracks chairs and supervised by the volumes editors through the Editorial Manager of the SUCI "Sustainable Civil Infrastructure" book series. As a result, 15 volumes have been formed of the final +320 accepted papers. The authors of the accepted papers have addressed all the comments of the reviewers to the satisfaction of the tracks chairs, the volumes editors, and the proceedings editor. It is hoped that readers of this proceedings of the GeoMEast 2017 will be stimulated and inspired by the wide range of papers written by a distinguished group of national and international authors.

Publication of this quality of technical papers would not have been possible without the dedication and professionalism of the anonymous papers reviewers. The names of these reviewers appear in the acknowledgment that follows. For any additional reviewers whose names were inadvertently missed, we offer our sincere apologies.

We are thankful to Dr. Hany Farouk Shehata, Dr. Nabil Khelifi, Dr. Khalid M. ElZahaby, Dr. Mohamed F. Shehata, and to all the distinguished volumes editors of the proceedings of the GeoMEast 2017. Appreciation is extended to the authors and track chairs for their significant contributions. Thanks are also extended to Springer for their coordination and enthusiastic support to this conference. The editors acknowledge the assistance of Ms. Janet Sterritt-Brunner, Mr. Arulmurugan Venkatasalam in the final production of the 15 edited volumes "Proceedings of GeoMEast 2017".

Contents

Optimizing the Position of Twin Tunnels for Reduction of Surface Ground Movements

Mehdi Mokhberi(✉) and Seyed Ali Farjam

Department of Civil Engineering, Estahban Branch, Islamic Azad University, Estahban, Iran

Abstract. In this study the deformations, ground movements, and stability of tunnels have been considered. To evaluate the effect of the position of a tunnel on surface settlement, a numerical study of twin tunnels in Shiraz City has been carried out. According to the different types of soils and the tunnel positions, the ground deformation is analyzed and discussed. Three different configurations have been taken into consideration: side-by-side, piggyback and angular-offset. Field data have been used to calculate the ground settlements caused by twin bored tunnels. A non-linear finite element analysis was performed to evaluate the surface deformations. Using the superposition approach, the empirical data have been compared with the analytical results. The results confirm that there is a clear difference in the amount and distribution of ground movements with varying tunnel positions.

Keywords: Twin tunnel · Underground · Tunneling · Settlement · Ground movements

1 Introduction

Underground transportation facilities in urban areas have a significant impact on citizens' lives, as they provide fast and safe transportation. As a result, the use of metro tunnels worldwide has increased considerably. Constructing multiple tunnels presents increased engineering challenges, e.g., the excavation of twin metro tunnels side by side or piggyback requires much more attention than constructing a single tunnel. Ground surface settlement is an inevitable consequence of excavating and constructing tunnels (Addenbrooke and Potis 2001).

In order to evaluate the behavior of multiple tunnels, a number of studies have been carried out. These studies focused on experimental, empirical and numerical methods. Many semi-empirical equations have been introduced by various authors for predicting the movements above single tunnels (O'Reilly and New 1982; Peck 1969). The equations have been shown over many years, when compared to case history data of single tunnels, to make accurate predictions of the vertical and horizontal displacements. Furthermore, numerical modeling and in situ observations have been used to analyze the interaction between twin tunnels. Results show that in some configurations, the interaction could greatly affect the soil settlement and that the design of twin tunnels requires numerical analysis associated with monitoring during the design and construction (Chehade and Shahrour 2008). The construction of the first tunnel may significantly

S. Agaiby and P. Grasso (eds.), *Engineering Challenges for Sustainable Underground Use*, Sustainable Civil Infrastructures, DOI 10.1007/978-3-319-61636-0_1

affect the soil conditions: reduced confinement, stress release, and a reduction of the strength parameters of the soils. Consequently, the second tunnel is excavated through a different material and the induced settlements related to the second tunnel will generally be greater (Guglielmetti et al. 2007). Many twin tunnel case histories do not have sufficient data points in critical positions and are usually given as a total profile and do not show the individual contributions made by each tunnel. Quality data that does fit this description has been reported (Cooper et al. 2002; Nyren 1998). There is a lack of data, but through advancements in computing power, researchers have been aiding the understanding of deformation of the above-mentioned twin tunnels using numerical methods (Addenbrooke and Potis 2001; Chehade and Shahrour 2008; Cooper et al. 2002; Chapman et al. 2003). Unfortunately, the magnitudes of the settlements predicted by using these numerical methods do not accurately reproduce the actual magnitudes of the displacements found at full scale. Hence, they cannot be used directly to improve the empirical equations. For this reason, this paper presents a 3D numerical analysis conducted to investigate the influence of twin tunnel spacing on the surface settlement and internal forces resulting from the tunnel excavation. The analysis was carried out for three different tunnel situations and different twin tunnels spaces.

2 Materials and Methods

In this study the analysis of the three types of twin tunnel configuration has been considered: side-by-side, piggyback and angular-offset. Figure 1 illustrates the tunnels' status. Common practice was carried out to superimpose the independent settlements to get the final settlement profiles. Using a finite element program, the soil was analyzed to evaluate the ground surface deformation. This research was performed to evaluate the tunnels' behavior, when constructed in the different classes of soils.

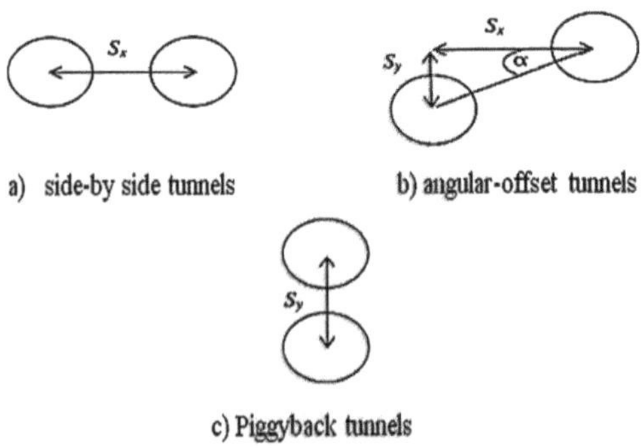

Fig. 1. Schematic of the position of twin tunnels

2.1 Soil and Lining Properties

In order to use the PLAXIS software, the properties of the soils and lining were defined. Five types of Shiraz soil properties and 35-cm thick concrete fragments have

Table 1. The data collected from different locations of Shiraz metro soil

	Soil classification	Location	Unit weight (γ) (kN/m^3)	SPT	W %	γ_{sat} (kN/m^3)	E (kN/m^2)	φ (degree)	C (kN/m^2)
1	GM-GP	Moa'li Abad	21	50	5	22	60000	40	5
2	GP	Qasr Dasht	19	15	15	20	40000	30	10
3	SM-SC	Farhang Shahr	18	40	10	19	25000	25	15
4	CL-ML	Khak Shenasi	17	12	12	19	30000	20	20
5	CL	Moshir Fatemi	20	25	15	22	20000	10	30

Table 2. Properties of concrete lining fragments

Paramerrs	Amount	Unit
EA	1.4×10^7	kN/m
EI	1.43×1^5	kNm/m^2
D	0.35	m
w	8.4	kN/m/m
υ	0.15	-

been used. The Mohr-Columb elasto-plastic criterion introduces the soil behavior. The data used in the analysis are listed in Tables 1 and 2.

2.2 Surface Effect

In order to determine the twin tunnel settlements, the surface deformation for each tunnel was found separately. An analysis was performed for the twin tunnels and the results measured. The obtained results are discussed below.

2.3 Side-by-Side Tunnel

Figures 2, 3, 4, 5, 6 and 7 show the effect of distance on the subsurface settlements. The 6 m diameter tunnels are buried in a depth of 15 m. The tunnels' spacing are 8, 16, 20, 24, 32, 40 and 60 m respectively. The water table position is 2 m above the tunnel lining. The settlements are dependent on the tunnels' spacing. The closest tunnels have more effect and larger settlements. Kept away from each other, the settlement decreases. The maximum settlement occurred at a distance 8 m from the center of each tunnel. The settlement remains constant between 8 to 16 m and decreases after 16 m from the center of the tunnels. At a distance of 60 m, the tunnels' settlements have no superimposed effects.

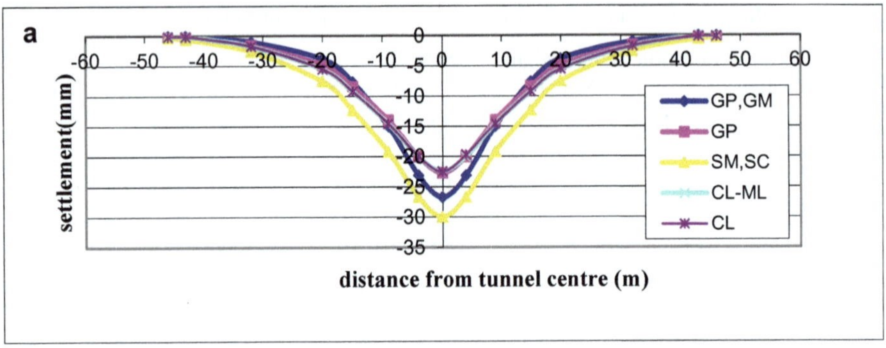

Fig. 2. The total settlement in side-by-side tunnels, tunnel diameter = 6 m; tunnel spacing = 8 m

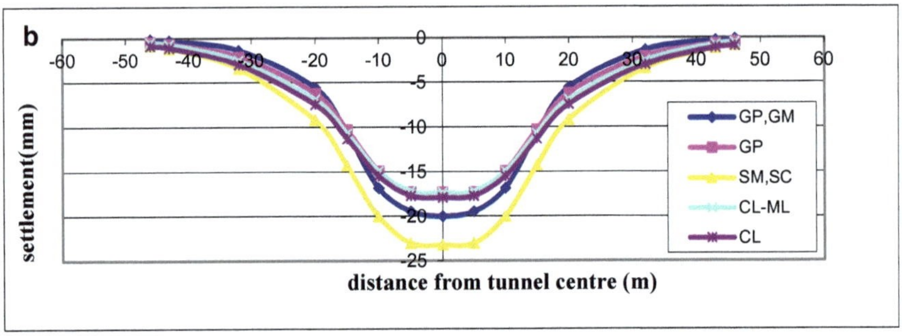

Fig. 3. The total settlement in side-by-side tunnels, tunnel diameter = 6 m; tunnel spacing = 16 m

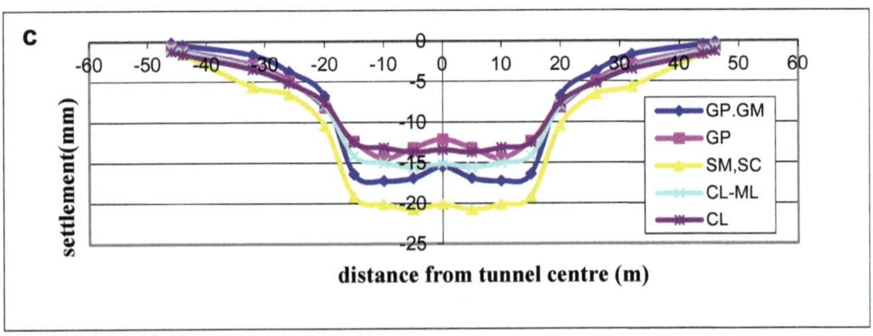

Fig. 4. The total settlement in side-by-side tunnels, tunnel diameter = 6 m; tunnel spacing = 20 m

2.4 Piggyback Tunnels

The Piggyback tunnels were analyzed at 8, 16, 20, 24, 32, 40 and 60 m distances. The water table position is 2 m above the tunnel lining. Figures 8 and 9 show the effect of

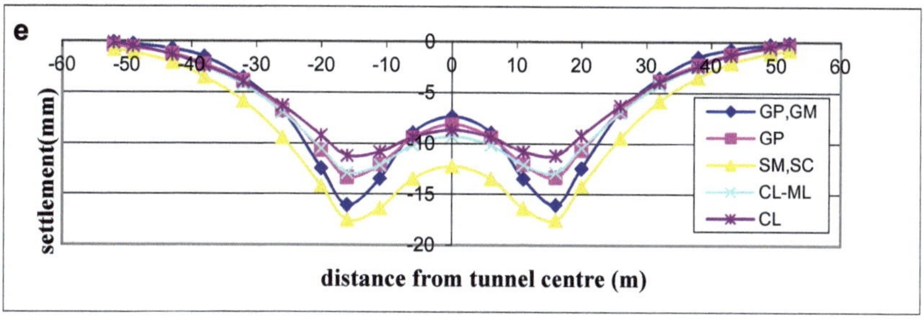

Fig. 5. The total settlement in side-by-side tunnels, tunnel diameter = 6 m; tunnel spacing = 32 m

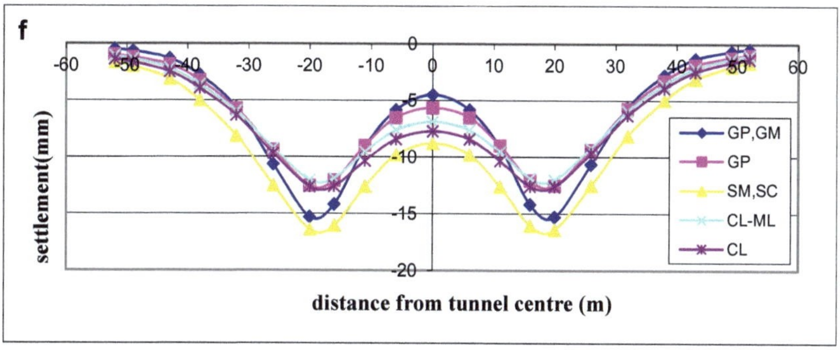

Fig. 6. The total settlement in side-by-side tunnels, tunnel diameter = 6 m; tunnel spacing = 40 m

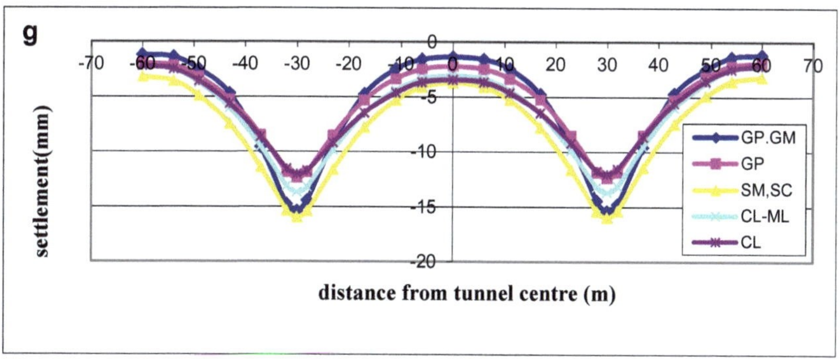

Fig. 7. The total settlement in side-by-side tunnels, tunnel diameter = 6 m; tunnel spacing = 60 m

vertical distance on the subsurface settlements. The results indicate that the piggyback tunnels have not interfaced significantly with the maximum settlements. The surface settlement is considerably dependent on the upper tunnel settlements.

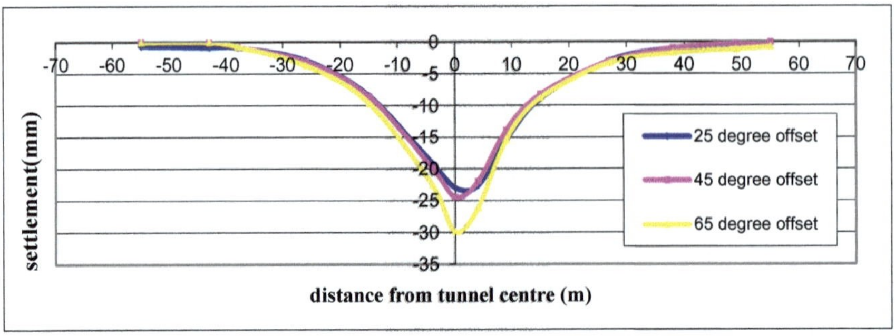

Fig. 8. Total settlement in angular-offset tunnels, tunnel diameter = 6 m; spacing = 8 m

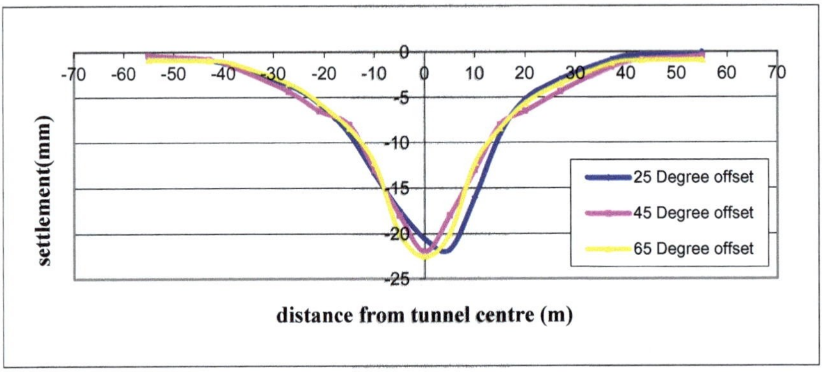

Fig. 9. Total settlement in angular-offset tunnels, tunnel diameter = 6 m; spacing = 14 m

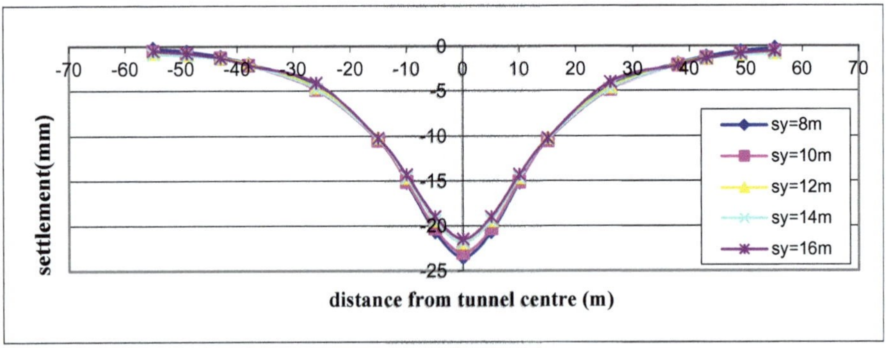

Fig. 10. The settlement in piggyback tunnel, tunnel diameter = 6 m

2.5 Angular-Offset Tunnels

For the angular-offset tunnels, an analysis of the cases of 25, 45 and 65° offset-angular has been carried out. The tunnels have 8 and 14 m offset distances and the soil class is CL.

The results show a linear decay with the spacing between the two tunnels. Figure 10 confirms that the settlements increase from a 25 to 65° offset. The 25° twin tunnel has less settlement and the 65° offset has the maximum superimposed settlements. The maximum amount of settlement has occurred in the center point of the twin tunnels.

3 Conclusions

The research was carried out to determine the effect of the situation of a twin tunnel on ground surface settlement. Using a finite element method analysis, the following results were obtained:

- In the single tunnel, settlements decrease with increasing the distance from the tunnel center; while the settlement has a maximum amount 8 to 16 m from the center of each tunnel in the twin tunnels, due to superimposed effects.
- In spite of soil saturation, the sand layers have considerable settlement compared with the other soil types.
- Settlement increases with increasing the offset-angular degree, and has the maximum amount in 60° alignments.
- The piggyback tunnels' settlement is due to the upper section's settlements: the lower parts have the least effect on superimposed settlements.

Acknowledgements. This paper was funded by a part of the M.Sc. Research in the Islamic Azad University, Estahban Branch. The authors are grateful to the head of the University, research management and the head of the civil engineering faculty. We also wish to thank the Shiraz Urban Railway Organization for their cooperation in providing the geotechnical data.

References

Addenbrooke, T.I., Potis, D.M.: Twin tunnel interaction: surface and subsurface effects. Int. J. Geomech. **1**(2), 249–271 (2001)

O'Reilly, M.P., New, B.M.: Settlements above tunnels in the United Kingdom - their magnitude and prediction. In: Proceedings of Tunneling 1982 Symposium, pp. 173–181. Institution of Mining and Metallurgy, London (1982)

Peck, R.: Deep excavation and Tunneling in soft ground. In: Proceedings of the 7th International Conference Soil Mechanics and Found Engineering, Mexico, pp. 225–231 (1969)

Chehade, F., Shahrour, I.: Numerical analysis of the interaction between twin-tunnels: influence of the relative position and construction procedure. Tunn. Undergr. Space Technol. **23**, 210–214 (2008)

Guglielmetti, V., Piergiorgio, G., Mahtab, A., Xu, S.: Mechanized Tunneling in Urban Areas. Taylor and Francis Group, London (2007)

Cooper, M.L., Chapman, D.N., Rogers, C.D.F.: Prediction of settlement in an Existing Tunnel cause by the Second of Twin Tunnels. Trans. Res. Rec. **1814**, 103–112 (2002). Design of structures

Nyren, R.: Field measurements above twin tunnels in London Clay. Ph.D. thesis, Imperial College (1998)

Chapman, D.N., Rogers, C.D.F., Hunt, D.V.L.: Investigating the settlement above closely spaced multiple tunnel constructions in soft ground. In: Proceedings of World Tunnel Congress 2003, Amsterdam 2003, vol. 2, pp. 629–635 (2003)

Research on the Opportunity of Construction Method Conversion in Upper-Soft and Lower-Hard Stratum Based on Pressure Arch Theory

Wei Wang(✉), Hengwen Zhang, and Jiasheng Bian

School of Civil Engineering, Central South University, Changsha 410075, Hunan, China
wangweicsu@csu.edu.cn, 345461777@qq.com, 2634105455@qq.com

Abstract. In order to determine the construction method conversion opportunity from Cross Diaphragm with Step method to three-bench method of Zi-Zhi Tunnel, this paper takes the shallow buried excavation section of the tunnel as an example, where the stratum is upper-soft and lower-hard. Considering different intersection positions between tunnel face and the interface of stratum, based on the pressure arch theory, the change laws for the pressure arch thickness under different conditions were analyzed by $FLAC^{3D}$. Then, a construction method conversion opportunity is derived. Meanwhile, the rationality of conversion opportunity is validated by the field monitoring. The results indicate that: the pressure arch thickness reaches a stable state and self-stable ability of surrounding rock reaches its maximum when the distance between tunnel bottom and the interface of strata is greater than 13 m. Hence, it is suggested to convert the construction method when the distance is greater than 13 m; the stable value and the convergence rate of vault subsidence are both below warning value in site, which can meet the safety and economic benefits of engineering.

Keywords: Pressure arch · Upper-soft and lower-hard · Construction method conversion

1 Introduction

Tunnel construction is mostly conducted in a single stratum. However, when tunnel passes through upper-soft and lower-hard formation, single excavation method could not meet the requirements of safe and schedule. In this case, excavation method should be adjusted dynamically according to the geological condition. Reasonable construction method conversion opportunity could reduce the disturbance of surrounding rock. Therefore, it is of significant value to study on the construction method conversion opportunity in upper-soft and lower-hard stratum.

Currently, the study on tunnel construction in upper-soft and lower-hard formation focused more on excavation methods comparisons (Diao and Li 2007; Zhang et al. 2011)

S. Agaiby and P. Grasso (eds.), *Engineering Challenges for Sustainable Underground Use*, Sustainable Civil Infrastructures, DOI 10.1007/978-3-319-61636-0_2

and safety research (Liu et al. 2014), while less on the influence of conversion opportunity on surrounding rock stability. Moreover, in construction practices, the conversion opportunity is mostly based on engineering experiences, though calculation method of relaxation pressure of shallow large span tunnel in up-soft and low-hard stratum has been obtained (Wang et al. 2014).

In order to determine the construction method conversion opportunity from Cross Diaphragm with Step method to three-bench method of Zi-Zhi Tunnel, based on the upper-soft and lower-hard stratum characteristic and the pressure arch theory, this paper reveals the change laws of the pressure arch thickness in different intersection positions between tunnel bottom and the interface of upper-soft and lower-hard. Engineering practice has verified the rationality of the opportunity. The results can provide references for future similar projects.

1.1 Engineering Survey

The length of Zi-Zhi Tunnel is 14.4 km. It is the longest urban tunnel of Asia. The project is divided into three sections: U-shape section, open cut section and shallow buried excavation section. This paper focuses on the shallow buried excavation section whose length, height and width is 877.1 m, 9.7 m and 12.8 m, respectively. The stratum is upper-soft and lower-hard, and the longitudinal profile is shown in Fig. 1. Design excavation methods of this shallow buried excavation section are Cross Diaphragm with Step method and three-bench method. Corresponding mileages of the two methods are shown in Fig. 1.

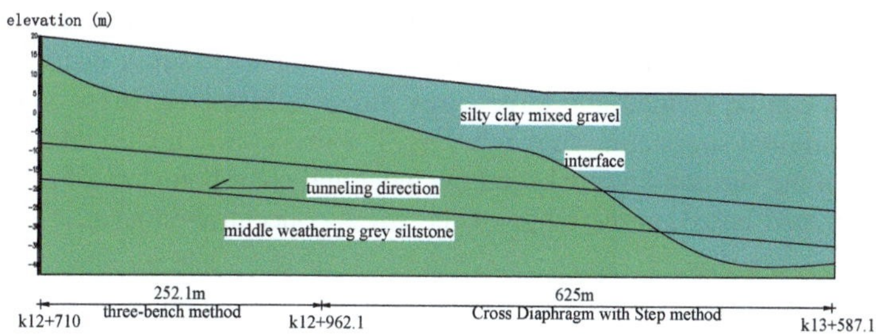

Fig. 1. The longitudinal profile of geology

With tunnel face moving forward, middle weathering grey siltstone (hereinafter referred to as the "bedrock") gets into the tunnel face and the invading bedrock depth of tunnel bottom is increasing. The excavation sequences and designed support parameters of Cross Diaphragm with Step method are shown in Fig. 2 and Table 1, respectively. In Cross Diaphragm with Step method, working face is divided into several smaller parts, which leads to repeated rock disturbances. In addition, the working space is so limited that only artificial and light equipments can be adopted. It is difficult to

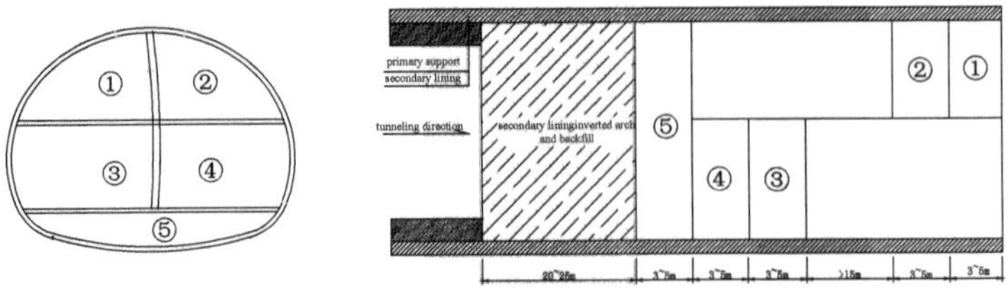

Fig. 2. Excavation sequences of Cross Diaphragm with Step method

Table 1. Support types and parameters of Cross Diaphragm with Step method

Support types	Support parameters
Primary support	C25 thick: 30 cm; I22 steel arch space: 50 cm
Secondary lining	C35 thick: 60 cm

ensure the progress of the project. To reduce the rock disturbance and ensure the construction period, construction method conversion from Cross Diaphragm with Step method to three-bench method should be conducted ahead of schedule.

1.2 Approach of Method Conversion Opportunity

1.2.1 Pressure Arch Theory

With the excavation of tunnel, initial stress field of surrounding rock is broken. And a free surface will apparent in tunnel face. Stress redistribution is produced to resist surrounding rock deformation. This phenomenon is the pressure arch effect (Lunardi 2000). According to the characteristics of surrounding rock stress distribution, surrounding rock can be divided into three cases: stress release area, stress concentration area and the original stress area, shown in Fig. 3. Stress concentration area is the main bearing unit of the surrounding rock, and the inner and outer boundary of pressure arch can be determined based on the range of the stress concentration area.

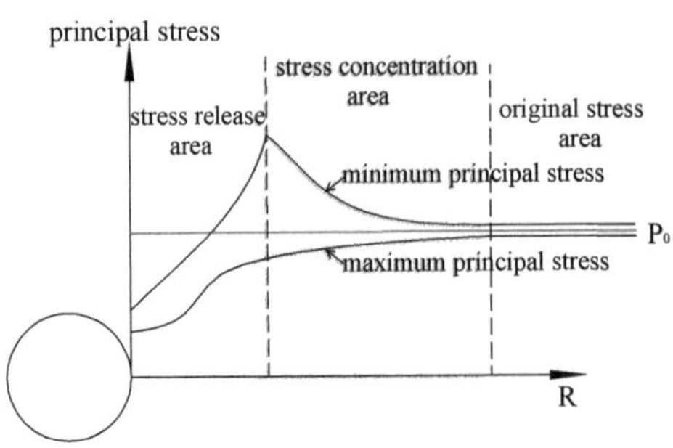

Fig. 3. Stress distribution of surrounding rock

(1) Determining method of internal boundary

The internal boundary of the pressure arch lies in tunnel perimeter if there is no horizontal stress decreasing area. Otherwise, the internal boundary lies in horizontal stress invariant points.

(2) Determining method of external boundary

Above the crown, the distribution of minimum principal stress is horizontal within pressure arch and it becomes vertical beyond pressure arch. Therefore, the direction changing point of minimum principal stress is taken as external boundary of pressure arch.

1.2.2 Calculation Model and Condition Design

Based on the geology condition, the stratum can be simplified, shown in Fig. 4. According to the geological survey and geological drilling, the calculation parameters of major soil are shown in Table 2. A 3D numerical model is established by $FLAC^{3D}$ and its horizontal length is 90 m, vertical 60 m, and the longitudinal 1 m, shown in Fig. 5. Solid element is used to simulate surrounding rock. In addition, rock and soil model adopt the Mohr-Coulomb constitutive model.

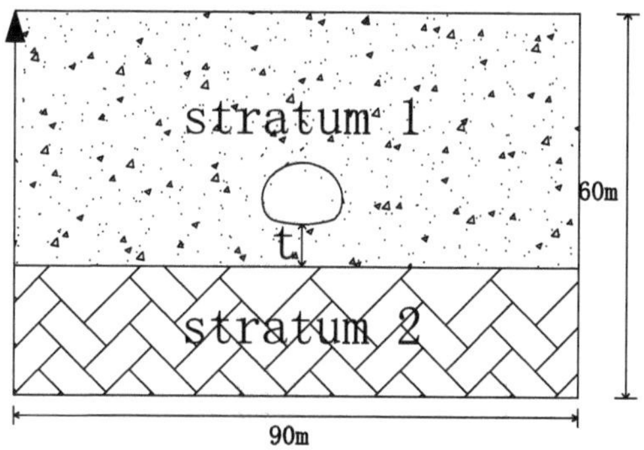

Fig. 4. Stratum simplified picture

Table 2. Calculation parameters in different stratum

Stratum	Density (kg/m^3)	Elastic modulus (GPa)	Poisson ratio	Internal friction angle (°)	Cohesion (KPa)
1	1980	0.2	0.29	17	45
2	2320	0.2	0.23	36	200

Note: 1-Silty Clay mixed Gravel, 2-Middle weathering Grey Siltstone.

In order to figure out the arching ability of surrounding rock in different intersection positions, assume that "*t*" refers to the distance between the interface of strata and tunnel bottom. "*t*" is negative when the interface is below the tunnel bottom, otherwise, "*t*" is positive. Calculation conditions are shown in Table 3.

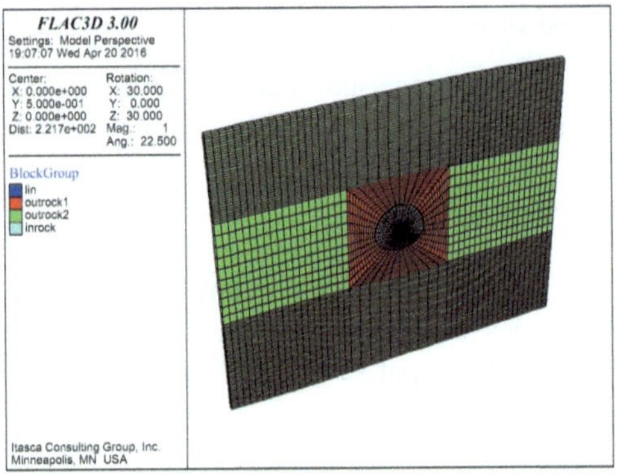

Fig. 5. Numerical calculation model

Table 3. Design of calculation conditions

Calculation conditions	1	2	3	4	5	6	7	8	9	10	11	12	13	14	15	16	17	18	19	20
t/m	−4	−3	−2	−1	0	1	2	3	4	5	6	7	8	9	10	11	12	13	14	15

1.2.3 Changing Laws of Pressure Arch Thickness

In order to analyze surrounding rock stress distribution after excavation, a typical stress monitoring path is defined. The path is from crown to the upper surface of model. Horizontal stress curves in monitoring path of No. 8 calculation condition are shown in Fig. 6 and horizontal and vertical stress curves are shown in Fig. 7.

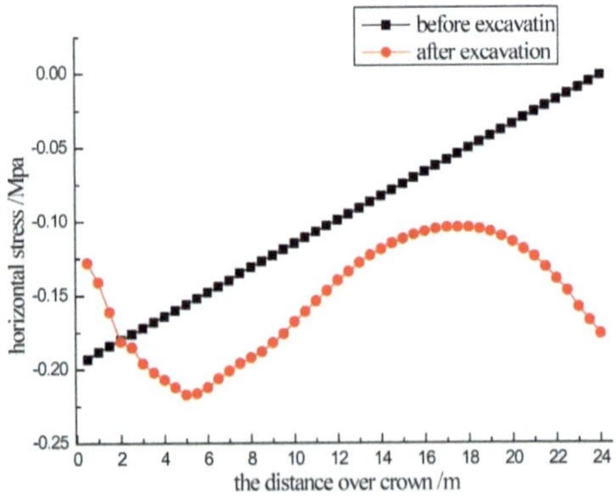

Fig. 6. Horizontal stress curve of No. 8 condition

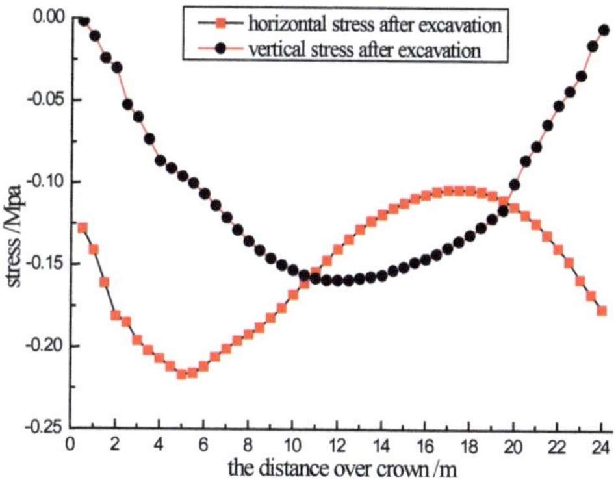

Fig. 7. Horizontal and vertical stress curves of No. 8 condition

According to the boundary determining methods of pressure arch mentioned, the intersection point of horizontal stress before and after the excavation is the internal boundary of pressure arch. And the first intersection point of horizontal and vertical stress after excavation is the external boundary. Figures 6 and 7 show that the distance between pressure arch internal boundary and crown, "a", is 2 m. And the distance between pressure arch external boundary and crown, "b", is 10.5 m. By that analogy, the statistics of pressure arch boundary about all calculation conditions are shown in Table 4 and the curve of pressure arch thickness is shown in Fig. 8.

Table 4 and Fig. 8 show that with the interface of strata increasing gradually, the changing laws of pressure arch thickness could be divided into three stages: no-arching stage, slow-decreasing stage and nearly-stable stage.

No-arching stage: when -4 m < "t" < 0 m, namely, before the bedrock invades tunnel bottom, pressure arch external boundary is confined to the ground surface, and could not form a natural pressure arch. The result indicates that stress concentration is serious and the arching ability of surrounding rock is insufficient, which is prone to cause landslides and large deformation.

Slow-decreasing stage: when 0 m < "t" < 13 m, the thickness of the pressure arch decreases quickly. The thickness of pressure arch above the vault represents the disturbance degree of surrounding rock and the self-bearing capacity. The smaller the thickness is, the less rock is required to bear load. The result indicates that with the invading bedrock depth of the tunnel bottom increasing, the arching ability of surrounding rock is improving.

Nearly-stable stage: when 13 m < "t" < 15 m, the curve is parallel to the "X" axis. The pressure arch thickness above crown tends to be stable and self-stable ability of surrounding rock reaches its maximum.

According to above analysis, it can be concluded that when the invading bedrock depth of the tunnel bottom is greater than 13 m, arching ability of surrounding rock reaches its maximum and it is conducive to utilize the self-bearing capacity of the rock mass. Therefore, it is suggested to convert the construction method after the distance between tunnel bottom and the interface of upper-soft and lower-hard stratum is greater than 13 m.

Table 4. Statistics of pressure arch boundary

t/m	a/m	b/m	The thicknesses of pressure arch/m	Produce pressure arch?
−4	2	/	∞	No
−3	2	/	∞	No
−2	2	/	∞	No
−1	2	/	∞	No
0	2	16.6	14.6	Yes
1	2	16.1	14.1	Yes
2	2	15.4	13.4	Yes
3	2	14.8	12.8	Yes
4	2	14.2	12.2	Yes
5	2	13.4	11.4	Yes
6	2	12.5	10.5	Yes
7	2	11.7	9.7	Yes
8	2	11.2	9.2	Yes
9	2	10.8	8.8	Yes
10	2	10.3	8.3	Yes
11	2	10.0	8.0	Yes
12	2	9.6	7.6	Yes
13	2	9.5	7.5	Yes
14	2	9.5	7.5	Yes
15	2	9.4	7.4	Yes

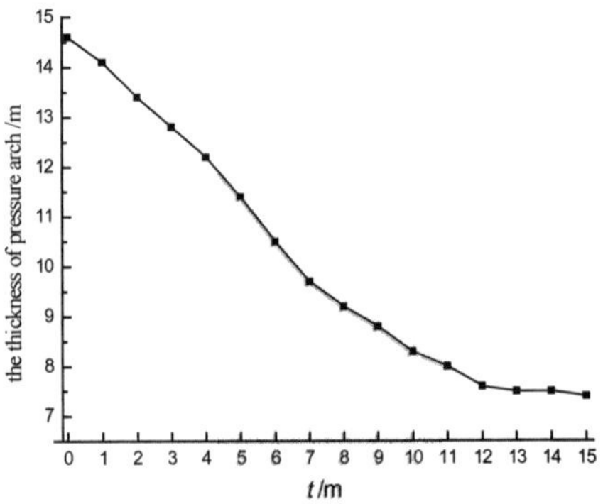

Fig. 8. Curve of pressure arch thickness under different conditions

1.3 Validation of Construction Method Conversion Opportunity

According to the construction method conversion opportunity above mentioned, the project department converted the Cross Diaphragm with Step method to three-bench method at K13 + 198.6. Excavation methods corresponding mileages in site are shown in Table 5.

Table 5. Tunnel excavation methods corresponding mileage in site

Excavation methods	Cross Diaphragm with Step method	Three-bench method
Start mileage	K13 + 587.1	K13 + 198.6
Terminal mileage	K13 + 198.6	K12 + 710
Length/m	388.5	488.6

Crown settlement and its settlement rate in K13 + 198.6 are measured. Deformation curve and deformation rate curve of crown settlement are shown in Figs. 9 and 10, respectably.

Figures 9 and 10 indicate that cumulative deformation increases rapidly and the crown settlement rate is large from an early time; After 15 days, the crown settlement rate slows down and cumulative deformation tends to be stable. The stable value of vault subsidence is 23 mm, which is lower than warning value, 30 mm. The rate of

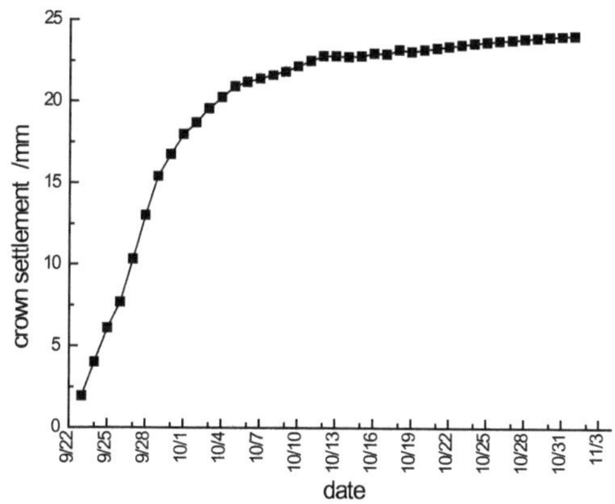

Fig. 9. Deformation curve of crown settlement

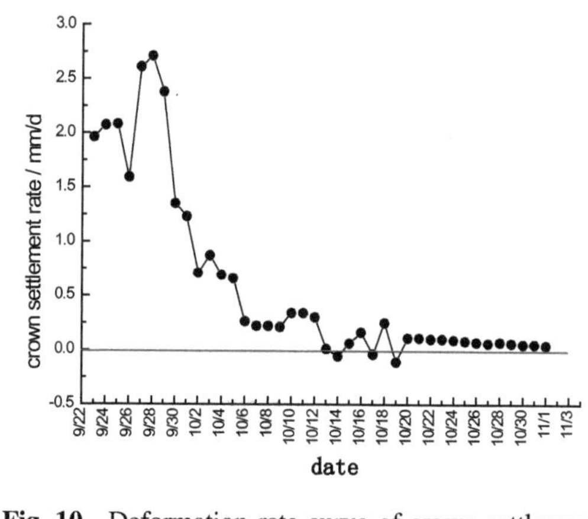

Fig. 10. Deformation rate curve of crown settlement

vault subsidence stable value is 0 mm/d, which is also lower than warning value 1 mm/d. The results show that the opportunity of construction method conversion can meet the safety benefit.

2 Conclusions

(1) In the shallow buried excavation section of Zi-Zhi Tunnel, stratum is generally upper-soft and lower-hard. To reduce the rock disturbance and ensure the construction progress, construction method conversion from Cross Diaphragm with Step method to three-bench method should be conducted ahead of schedule.
(2) According to geographical character of shallow buried excavation section of Zi-Zhi Tunnel that the invading bedrock depth of the tunnel bottom is increasing, the pressure arch thicknesses change laws under different conditions could be divided into three stages: no-arching stage, slow-decreasing stage and nearly-stable stage. Arching ability of surrounding rock reaches its maximum when the invading depth of tunnel bottom is greater than 13 m. Therefore, it is suggested to convert the construction method when the distance between tunnel bottom and the interface of strata is greater than 13 m.
(3) The stable value and its convergence rate of vault subsidence in the actual conversion mileage are both below warning value, which can meet the safe and economic benefits of engineering.

Acknowledgments. The authors appreciate the support of Science and Technology Plan Projects of Hunan Province, China (2010GK3173).

References

Diao, Z.G., Li, C.J.: Study on construction method of large-section tunnels in upper soft and lower hard layers. Tunn. Constr. **S2**, 433–436 (2007)
Zhang, J.X., Guo, F.C., Lei, G.: Study on construction methods of shallow-buried tunnel under condition of upper-soft lower-hard ground by numerical simulation. Chin. J. Undergr. Space Eng. **S1**(7), 1416–1421 (2011)
Liu, T., Zhao, S.Z., Sun, F.F.: Security analysis of dismantling temporary supports of large-span shallow tunnel in upper-soft and lower-hard stratum. Rock Soil Mech. **S1**, 306–310 (2014)
Wang, Z.W., Qiao, C.S., Song, C.Y.: Calculation method of relaxation pressure of shallow large span tunnel in up-soft/low-hard rock stratum. Rock Mech. **35**(8), 2342–2352 (2014)
Lunardi, P.: The design and construction of tunnels using the approach based on the analysis of controlled deformation in rocks and soils. T&T Int. ADECO-RS Approach (2000)

Computational and Numerical Considerations of Segmental Lining Design

Mostafa Zaki Abd Elrehim[1(✉)] and Mostafa Asaad[2]

[1] Civil Engineering Department, Faculty of Engineering Minia University, Minia 61111, Egypt
[2] Technical Department (TD), National Authority for Tunnels (NAT), Cairo, Egypt

Abstract. Thanks to the high safety level and the high-speed rate of work, mechanized tunneling (especially TBM) has speedily spread worldwide for tunnels excavation. Lining behind TBM normally consists of precast rings each one consists of a number of segments connected together with radial and longitudinal bolts. The main items which control the stability of TBM excavation are face pressure, grout pressure and interaction of lining components with grout and surrounding soil.

Many methods have been used to study the behavior of segmental lining including empirical methods, analytical methods, and numerical modelling. The third method is currently considered one of the most commonly used methods for design check purposes. Most tunnelling codes use analytical methods to study the behavior of segmental lining. The aim of this paper is to study the shortenings of current used numerical models and try to improve it, in addition, to check the efficiency of numerical modelling compared to analytical methods. Firstly a comprehensive study will be performed to get an appropriate model for simulating segmental lining components. Consequently, a better understanding will be possible for the interaction between different segmental lining components and the annular grout. The results of the analysis will be discussed and compared to the traditional models to check the efficiency and validity of the numerical modelling technique. Finally, a study will be introduced to investigate the optimum number of bolts may be used between successive segments. This could help in optimizing the segmental lining for less cost and more safety levels.

Keywords: TBM excavation · Lining components · Numerical modeling · Tunneling codes · Optimization

1 Introduction

Design of tunnels requires an adequate estimate of surface settlement and lining forces. The main principles which structural design need to satisfy are that tunnel should has sufficient strength and stiffness to resist all subjected loads, in addition, it should be stable during all the stages of construction and along its lifetime Bakker (2003). Many design methods have been used within this context, varying from simple empirical and analytical formulations to advanced finite element analyses. Many design models have

S. Agaiby and P. Grasso (eds.), *Engineering Challenges for Sustainable Underground Use*, Sustainable Civil Infrastructures, DOI 10.1007/978-3-319-61636-0_3

been developed by engineers and approved by different associations such as Japanese tunneling association, Federal highway administration (U.S.A) and British Tunneling Society, in addition, the ITA suggested a number of design models ITA (1988), ITA (2000).

Generally, all design methods can be classified into the following four major types Lee *et al.* (2001), Hu *et al.* (2009): (a) Empirical design methods based on past tunneling projects. (b) Design methods based on in situ measurement and laboratory testing. (c) Circular ring on elastic foundation method (d) Continuum mechanics models including analytical methods and numerical methods.

So far the third method is currently considered the most commonly adopted for design purpose Wood (2002) and Hu *et al.* (2009). This method can also be categorized into analytical methods and numerical methods.

Numerical methods can be categorized according to the treatment of the joints of segments to the following categories: Koyama (2000), Koyama (2003), JSCE (2007), Gruebl (2006), Hu *et al.* (2009) and Gruebl (2012) (Fig. 1).

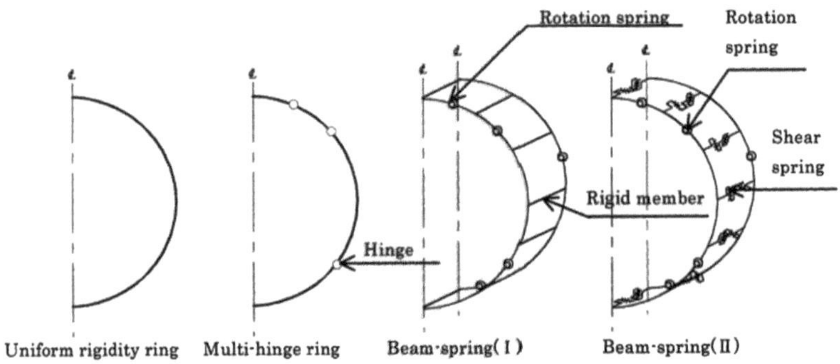

Fig. 1. Structural models of the segmental ring Koyama (2000)

1. Uniform rigidity ring
2. Multi-hinge ring
3. Beam-spring (I)
4. Beam-spring (II)

Nearly all models used nowadays to analyse segmental lining depend on the above categories to simulate the radial and longitudinal joints between segments/rings. All methods employ different assumptions to simulate the joints and the interaction between joints and segments. The aim of this study is to enhance existing numerical modelling for more understanding of the behavior of segmental lining and the interaction between its components and the surrounding grout and soil layers. This aim can be achieved by modelling lining components (segments and bolts) separately, and by using elements which can simulate the real behavior of the connecting bolts. Thus the effect of bolts in the stability of lining can be independently studied.

2 Soil Modelling

During the construction of the lining and before the hardening of the grout, there is no direct interaction between the lining and the soil Wood (2002), Blom (2003). To date and to the knowledge of the authors, this case is the most critical case during tunnel construction. Most failure cases happen during this stage. Therefore, it will be reasonable to model the confinement provided by the soil by means of springs. The focus of this paper is to study the local behaviour of the lining during construction. The more sophisticated modelling of the whole soil layers together with lining elements is not practically visible to accurately predict the local behaviour of the lining components.

Knowing that the studied case in this paper focuses on the stage where the grout doesn't got hardened yet and no direct contact between soil and lining, accordingly it will be useful to use springs for modelling the soil to focus on the behavior of the lining itself.

2.1 Stiffness of Springs

The confinement provided by the soil has been modelled using spring elements. These types of support resist lining movement in direct bearing and tangential shear across the grouted layer surrounding the lining. The springs' stiffness were calculated using linear load-deformation relationship according to Duddeck and Erdman (1982), JSCE (2007), Mahajan (2010) and Gruebl (2012) as follows:

$$E_c = \frac{E\,(1 - \nu)}{(1+\nu)(1-2\nu)}$$

$$C_r = \frac{Ec}{R}$$

$$K = Cr * A$$

Where:

E, ν = Young's Modulus and Poisson's Ratio of the ground.
R = equivalent tunnel radius.
A = area of soil that is to be represented by the equivalent radial spring. This is the distance between adjacent radial springs multiplied by a unit length.

3 Spring Constants Calculations

Spring constants can be determined based on tributary projections on the x and y-axis of each joint. If the analysis software being used supports the use of radial springs, then all spring constants will be the same. For programs which don't support this option (MIDAS GTS as an example) the following formulas can be used to determine spring constants in (X, Y) directions FHWA (2009) Figs. 2 and 3:

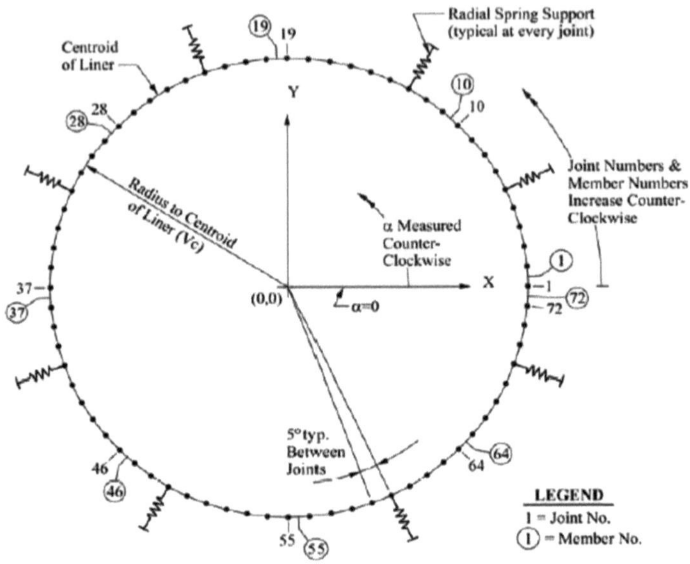

Fig. 2. Joints and members – computer model

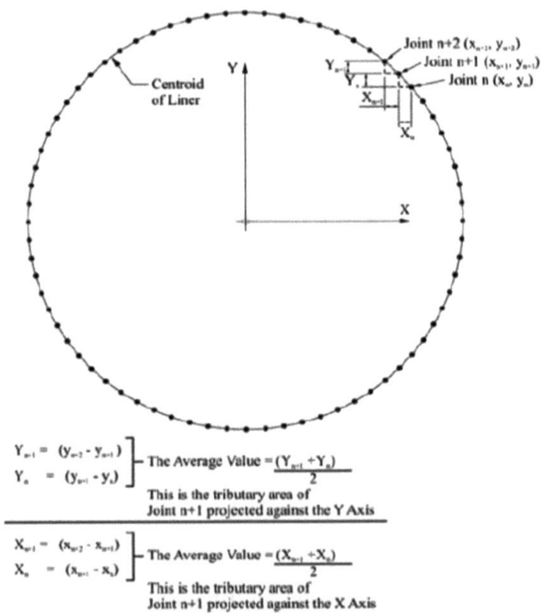

Fig. 3. Spring constant computation

Spring constant in Y direction = $K_s(X_n + X_{n+1})/2$

Where:

$$X_n = |(X_n - X_{n+1})|$$
$$X_{n+1} = |(X_{n+1} - X_{n+2})|$$

Spring constant in Y direction = $K_s(Y_n + Y_{n+1})/2$

Where:

$$Y_n = |(Y_n - Y_{n+1})|$$
$$Y_{n+1} = |(Y_{n+1} - Y_{n+2})|$$

Figures 2 and 3 illustrate the method for calculating orthogonal springs.

In the aforementioned equations:

The coordinates for joint N = (x_n, y_n)
The coordinates for joint N+1 = (x_{n+1}, y_{n+1})
The coordinates for joint N+2 = (x_{n+2}, y_{n+2})

Ground behaviour in the tangential direction is generally ignored in beam spring model method FHWA (2009).

4 Data Used in Modelling

The proposed model used in this study (as a reference model) was developed by Abu-Krisha (1998). It is a three-dimensional numerical model used to analyse TBM tunnelling Abu-Krisha (1998), in addition some other data got from National Authority of Tunnels at Cairo NAT (2009). Bolts properties values are calculated from reference paper *Hinchberger et al.* (2010).

Tables 1 and 2 show the data of rings and the properties of materials (Concrete and Grout) used in the model.

Table 1. Data of lining used in the modelling.

Lining data		
Item	Concrete lining	Grout
Outer diameter (m)	8.7	9
Thickness (m)	0.35	0.15
Length of ring (m)	1.5	–
E (MPa)	30000	0.1 to 20E3
ν	0.2	0.4 to 0.18

Table 2. Parameters of different materials used in the analysis.

Parameter	Segments (concrete)[a]	Grout[a]	Bolts[b]	Shield tail[a]
Initial tangent elastic modulus (E) MPa	30000	0.1 to 20 E3	210000	30E3
Poisson's ratio (ν)	0.2	0.4	0.3	0.3
Density kg/m^3	2600	–	8030	–
Unconfined compressive strength (fcu) Mpa	60	–	–	–
Initial yield stress (σy) MPa	48	–	350	–
Tensile failure stress MPa	4	–	800	–
Type of selected element	Plate element	Solid elem	Truss element	Plate element
Model behavior	Elastic	Mohr coulomb	Von mises	Elastic

[a]*Following the data from reference model Abu-Krisha* (1998).
[b]*Following the data from reference paper Hinchberger et al.* (2010). Bolts properties are same for segments and rings bolts.

5 Numerical Modelling

In this section, the suggested approach used to model the segmental lining components will be presented. The FEM software used in this study is Midas GTS (MIDAS GTS NX V3.0.0). Three dimensional simulation has been used to study the local behavior of TBM segmental lining. Consequently, it will be possible to accurately simulate and study the behaviour of each component of TBM segmental lining. In this technique, lining is modelled with shell elements, grout is modelled with solid elements and bolts (between segments and between rings) are modelled with truss elements (Figs. 4 and 5). Knowing that type of bolts which have been used (for both radial and tangential direction) is steel bolts with Elastic Modulus of 210000 MPa.

Each of the aforementioned elements has been used to simulate the actual properties and behaviour of the lining elements (Segments, joints, and grout) separately. Finally the studied case in this paper is for normal construction condition during TBM advance, at segments just behind the machine.

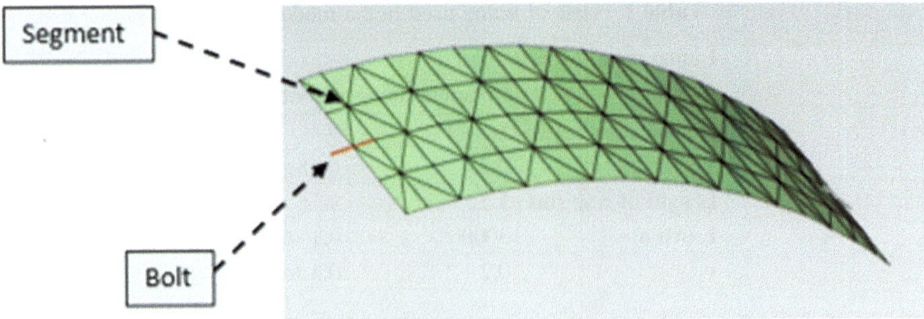

Fig. 4. Simulation joint for current developed model

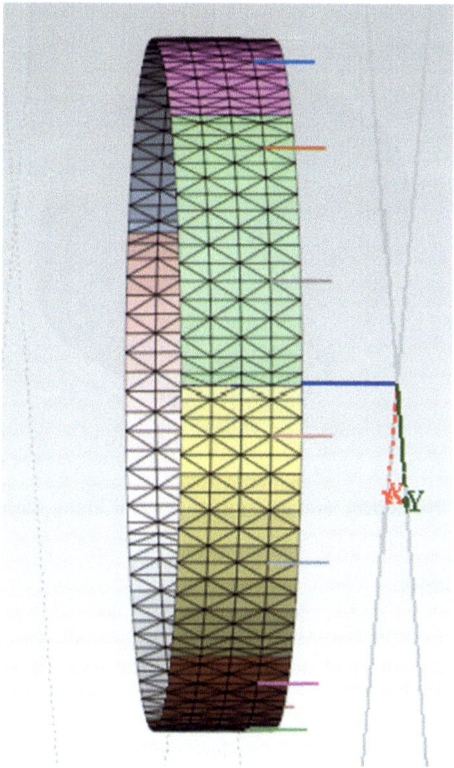

Fig. 5. Simulation of bolts connecting successive rings (one bolt between each two facing segment).

5.1 Boundary Conditions

As indicated in Fig. 6, three rings have been considered in the model. The considered ring will be the middle ring (number 2). Ring No. 1 (at the right) is already the installed ring before the considered one. Ring Number 3 (at the left) is still inside the shield tail of the machine. Boundary conditions for the model will be hinged support at the end of ring number 1 and support condition due to thrust load at ring number 3.

5.2 Load Cases

Load Case 1: Activation of ring No. 1 inside the shield.
Load Case 2: Activation of ring No. 2 inside the shield, activation of grout layer for ring No. 1.
Load Case 3: Activation of ring No. 3 inside the shield, activation of grout layer for ring No. 2.

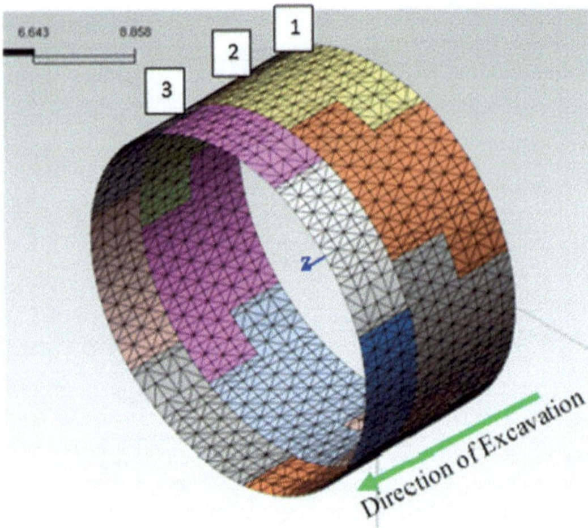

Fig. 6. Numerical simulation of segmental lining (case study)

5.3 Analysis of Results

Referring to Table 3, a comparison has been performed between the current developed model and the existing numerical models which used to simulate segmental lining.

Table 3. Comparison of results between different models

	Solid ring model	Muirwood model	BSM model	Current developed approach
Vertical deformation (mm)	−14	−14.8	−14.3	−15.4
% difference	−9.09%	−3.90%	−7.14%	–
N.F [kN/m]	−1075	−1062	−1067	−1999
% difference	−46.22%	−46.87%	−46.62%	–
B.M [kN.m/m]	−10.7	−9.7	−7.5	−7.5
% difference	39.87%	26.80%	−1.96%	–
Axial force at longitudinal bolts [kN]	–	–	–	−800
Axial force at radial bolts [kN]	–	–	–	−50

Calculation of result values from existing numerical models have done as followed:

Uniform rigidity ring:

Calculations have been performed following the calculation method according to Koyama (2003) JSCE (2007) and Hu et al. (2009).

Muirwood model:

Calculations have been performed following the calculation method according to Gruebl (2006) and Gruebl (2012).

Beam spring model (BSM):

Calculations have been performed following the calculation method according to Koyama (2003); JSCE (2007) and Hu et al. (2009)

Sample of the analysis results are shown in Figs. 7, 8, 9 and 10, the following points can be concluded:

- The crown deformation value from the current developed system is close to the crown deformation values from other models.
- Normal forces of the developed model are higher than the other models.
- Axial force for bolts can be directly calculated using this model, consequently, design and check of bolts section can be proceeded using this model.
- The value of axial force for radial bolts is very small compared to its value for longitudinal bolts. Consequently, the effect of longitudinal bolts in the stability of the ring is greater than radial bolts.
- So far and reffering to current technique results, the common used methods aproximations can be safely used in the design.
- Using the current technique, accurate design of bolts can be directly achieved.
- Although the current technique considered bolts and segmetns seperately, it's a simple model. It doesn't require long time or big effort to be achieved.
- Studying the behavior of segmental lining during construction is so vital. Using the developed technique, effect of bolts and annular grout on the stability of lining during construction can be studied. Thus, accidental cases during construction can be greatly avoided.

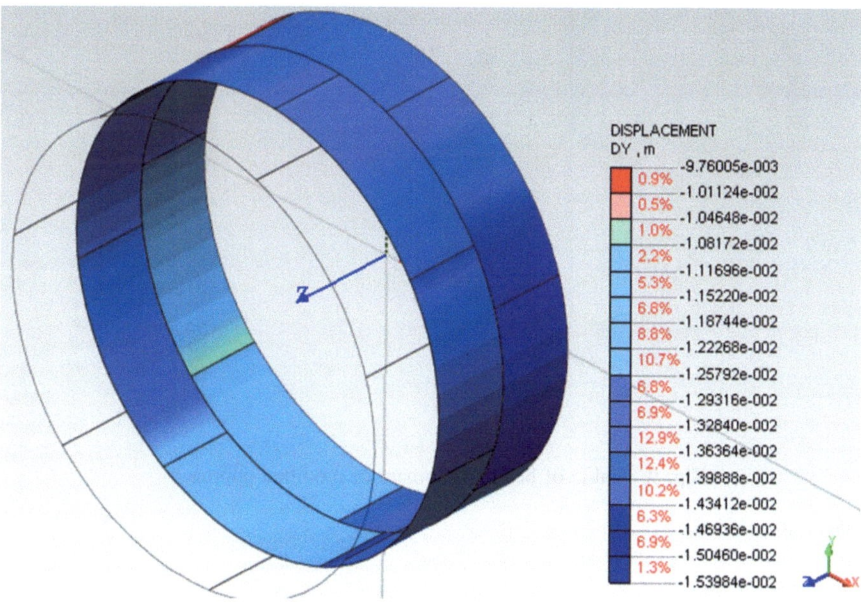

Fig. 7. Crown deformation after ring construction

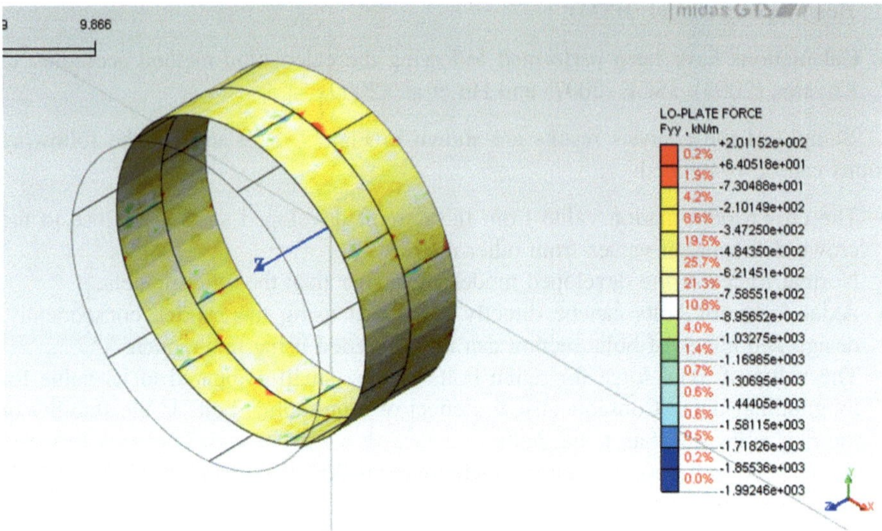

Fig. 8. Value of normal force of modified approach

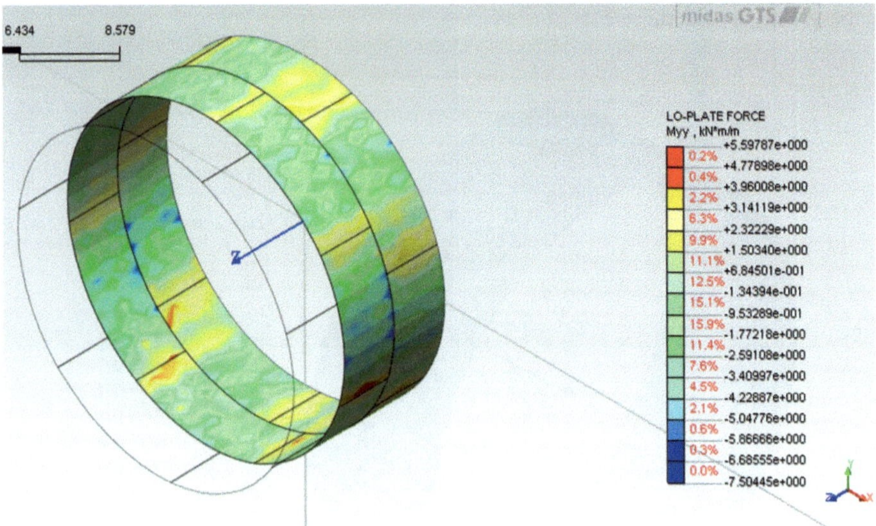

Fig. 9. Value of bending moment of modified approach.

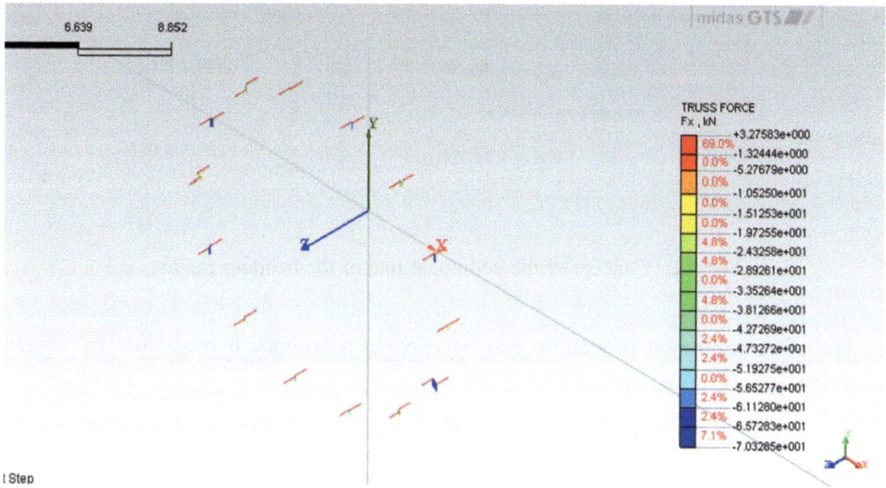

Fig. 10. Axial force for radial bolts

6 Current Technique Versus JSCE Equations Results

In the following part, values of bending moment of segmental lining developed according to Japanese Tunneling Association equations will be calculated and compared to bending moment values from current developed technique.

The effect of longitudinal joints and circumferential joints is simulated by a simple model recommended by Japanese Tunneling Association JSCE (2007). In this model, the deformation of a ring with joints is compared to that without any joints. The ratio between the two values is assumed to be η. Where η is the effective ratio of the bending rigidity (Fig. 11). In addition, a bending moment whose magnitude is additional rate ξ times that of the segment is considered to be distributed to the neighbor segment in the joint area (Fig. 12). At that point, the values of the bending moments used for the segment and the joint are assumed to be Mo (1 + ξ) and Mo (1 − ξ), respectively.

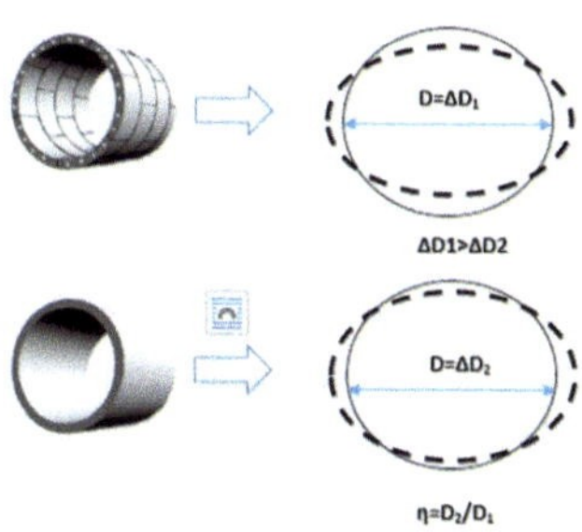

Fig. 11. Concept of the efficient ratio of the bending rigidity

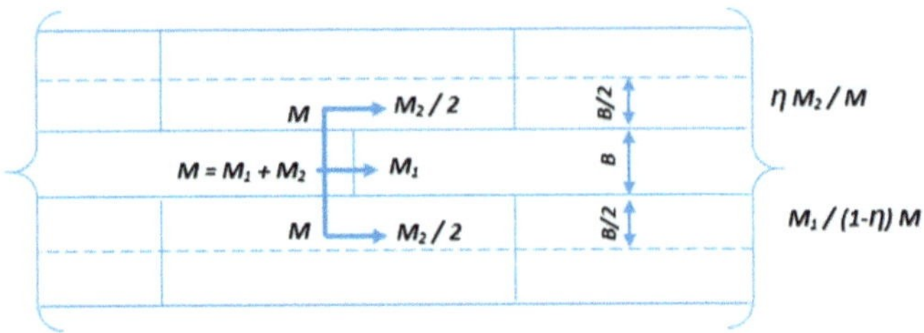

Fig. 12. Concept of the additional rate of the bending rigidity

- The aforementioned procedure can be used in a numerical procedure by several steps:

1. Correction of the elastic modulus of the ring according to a factor η.
2. Calculation of the stress characteristics.
3. Modification of the value of the bending moment, increasing and decreasing the value for the segment and the ring, respectively by ξ factor.
4. The normal force remains the same.

$$E_a = \eta . E_c \quad M_1 = (1-\xi).M \quad M_0 = (1+\xi).M$$

E_a = the virtual modulus of the ring
E_c = the concrete modulus
M = the bending moment calculated in ring with uniform flexural rigidity η. EI
M_1 = the bending moment of the joint
M_0 = the bending moment of the segment

Notice:

1. The effective bending ratio η is described as being universally determined by the profile of the joint and the shape and the size of the segment.
2. The value of the parameter ξ varies between 0.3 and 0.5 as a function of number of segments and the stiffness of the surrounding ground.

6.1 Calculations of Bending Moment for Current Example

The data of current example have been used to calculate the bending moment depending on the equations developed by JSCE. These values will be compared with values from the current developed model.

To calculate η value, a model has been considered with solid ring without any joints as mentioned earlier in this section. The crown deformation value equals to: −1.4 cm.

Also another model has been considered with hinged joints between segments. The crown deformation value equals to: −2.05 cm.

$$Ds = -1.4\ \text{cm}$$
$$Dj = -2.05\ \text{cm}$$
$$\eta = \frac{Ds}{Dj} = -1.4/-2.05 = 0.682927$$
$$\xi = .4\ (\text{As medium diameter rings})$$

$$\eta = .682927 \quad E = 30000000\,\text{kN/m}^2 \quad \xi = .4$$
$$\text{Moment of Inertia for section} = 1 * .35^3/12 = .003573\,\text{m}^4$$
$$E_a = \eta.E_c = 0.682927 \times 30000000 = 20487804.88\,\text{kN/m}^2$$

Model with modified E has been achieved and crown Bending Moment value has been calculated

$$M = -6.2\,\text{kN.m/m}$$

Eventually values of M_0 and M_1 will be calculated as follows:

$$M_0 = (1 + \xi).M = (1 + .4) * -6.7 = -8.7\,\text{kN.m/m}$$
$$M_1 = (1 - \xi).M = (1 + .4) * -6.7 = -4.7\,\text{kN.m/m}$$

6.2 Comparison of Results

Notations:

JSCE Equations Results Approach: JA
Current Approach: CA

$$M_{max}\ \text{for (JA)} = -8.7 \quad M_{max}\ \text{for (CA)} = -7.5$$
$$M_{Joint}\ \text{for (JA)} = -4.7 \quad Mj_{oint}\ \text{for (CA)} = +3.72$$

$$\text{Crown deformation for JA} = -1.44 \quad \text{Crown deformation for CA} = -1.54.$$

The aforementioned results show that maximum moment is quite close for both approaches, which means that the assumptions of JSCE equations can be used in design. On the other hand values of joints bending moment from these equations is not accurate. Which means that due to its assumptions, traditional methods can't accurately interpret the distribution of moment between rings and joints. In addition, the inaccuracy of ξ value (.3 to .5) makes it hard to get an accurate value of straining actions and always need to be calibrated in each problem.

7 Estimation of Optimum Number of Connected Bolts

In the following section, a brief study will be demonstrated to investigate the optimum number of bolts to be used between segments (Fig. 13).

Model No. 1: one bolt for each joint between segments.
Model No. 2: two bolts for each joint between segments.
Model No. 3: three bolts for each joint between segments.

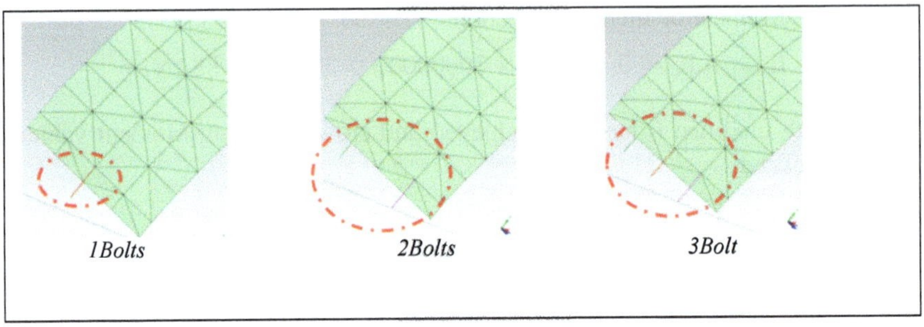

Fig. 13. Different systems used for comparison

7.1 Load Cases

Load Case 1: Activation of ring No. 1 inside the shield.
Load Case 2: Activation of ring No. 2 inside the shield, activation of grout layer for ring No. 1.
Load Case 3: Activation of ring No. 3 inside the shield, activation of grout layer for ring No. 2.

7.2 Analysis of Results

From the results shown in Table 4 and Figs. 14 and 15 the following notices can be concluded:

- Difference between deformations of the three models is small.
- The normal force value calculated from the 1-Bolt system is very high compared to the value from traditional methods.
- The normal force value from the 2-Bolts system is close to the normal force from traditional methods.
- Difference between normal force values of the developed method using one bolt and using two bolts and traditional methods shows that simulation of bolts and the number of used bolts are greatly affect the behavior of segmental lining.

- The normal force is greatly decreased with the increase of the number of bolts, difference between the normal force of the 1-Bolt system and the 2-Bolts system is greatly large. Whereas there is a slight difference between normal forces of the 2-Bolts and the 3-Bolts systems.
- The previous notice means that the assumptions of the traditional methods fit better when the used number of bolts at joints are equal or more than two bolts.
- This notice means that increasing the number of bolts between segments, makes the ring stiffer and makes its behaviour more close to the solid ring behaviour. This notice also explains the great effect of bolts in the stability of segmental lining during construction.

Therefore it can be concluded that; using 2-Bolts system is preferable and achieve optimum results in view with straining actions.

Table 4. Comparison of crown deformation and straining actions between different systems

	1 Bolt	2 Bolts	3 Bolts
Crown deformation (mm)	−15.48	−15.1	−15.1
Bending moment (kN.m/m)	+ 24	+ 5.5	+ 11
	−7	−7.4	−7.2
Normal force (kN/m)	−3765	−1993	−1410

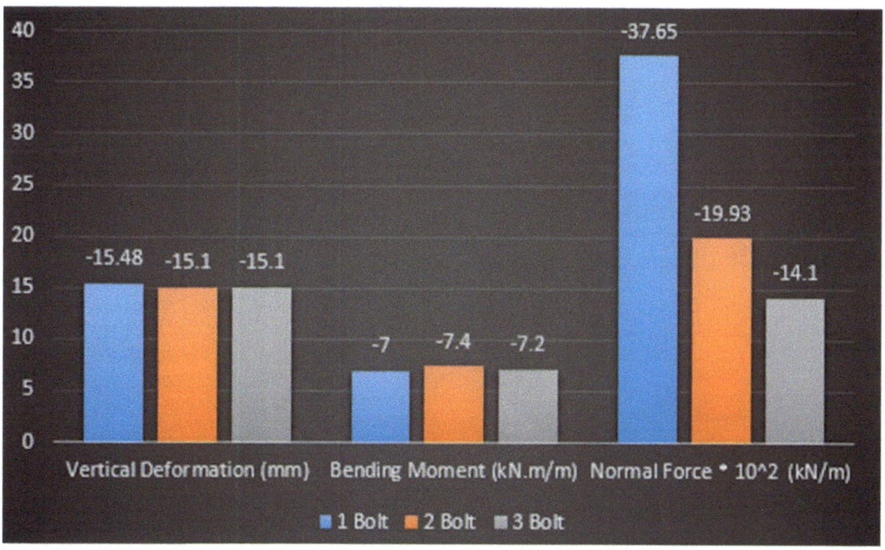

Fig. 14. Comparison of crown deformation and straining action between different systems

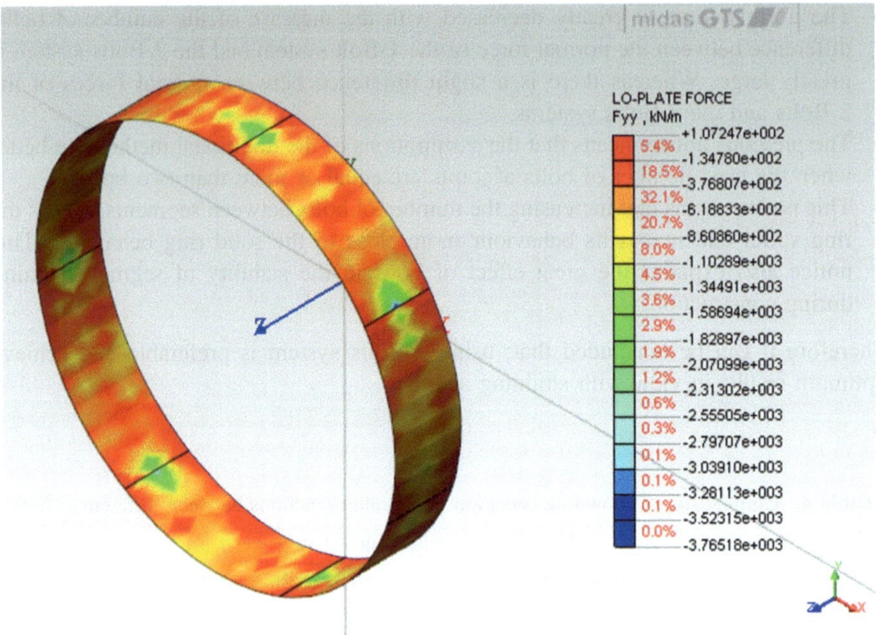

1 Bolt

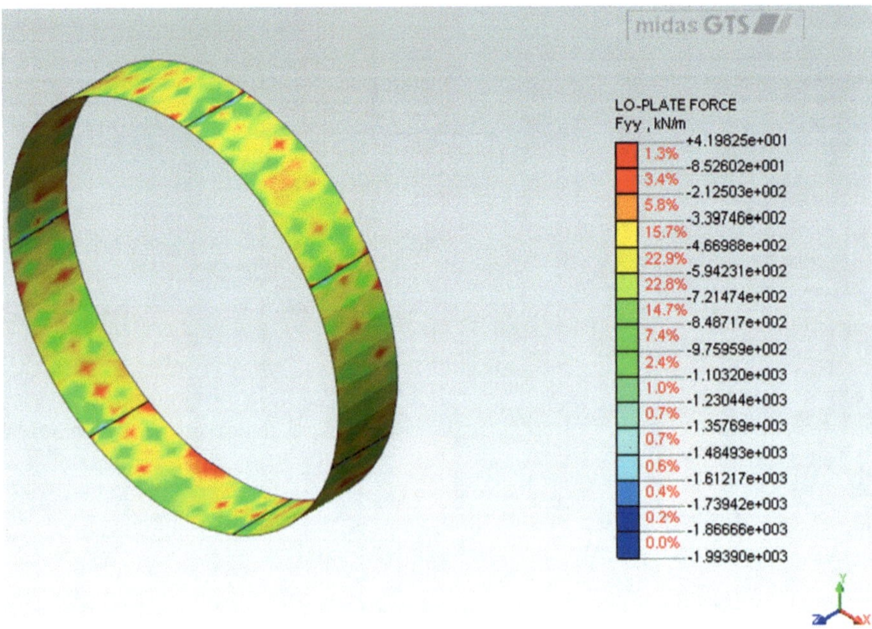

2Bolts

Fig. 15. Value of normal force for the three systems

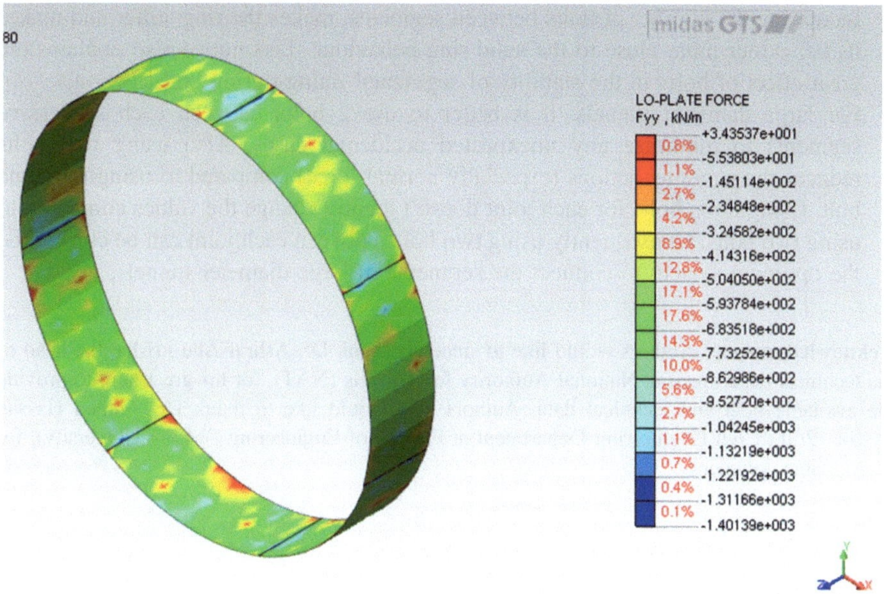

3 Bolts

Fig. 15. (continued)

8 Conclusions

From this study the following points can be concluded:

- So far and reffering to current technique results the common used methods aproximations can be safely used in the design.
- Studying the behavior of segmental lining during construction is important. Using the developed techniques, effect of bolts and annular grout on the stability of lining during construction can be studied. Thus, accidental cases during construction can be greatly avoided.
- The straining actions resulted from the new developed technique (using 1 bolt) are very high compared to using 2 bolts and to traditional methods.
- Axial force for bolts can be directly calculated using the new developed model, consequently, design and check of bolts section can be proceeded using this model.
- Simulation of bolts and the number of used bolts are greatly affect the behavior of segmental lining.
- The value of axial force for radial bolts is very small compared to its value for longitudinal bolts. Consequently, the effect of longitudinal bolts on the stability of the ring is greater than radial bolts.
- The assumptions of the traditional methods fit better when the used number of bolts at joints is equal or more than two bolts.

- Increasing the number of bolts between segments, makes the ring stiffer and makes its behaviour more close to the solid ring behaviour. This notice also explains the great effect of bolts in the stability of segmental lining during construction.
- For large diameter tunnels, it is better to use 2 bolts between each successive segments to minimize any unexpected accidental cases. Also using two bolts reduces the straining actions (especially normal force) compared to using only one bolt. Using three bolts for each joint doesn't greatly change the values compared to using two bolts. Consequently using two bolts between each joint can be considered the optimum system to connect the segments in large diameter tunnels.

Acknowledgments. Authors would like to sincerely thank Dr. Ashraf Abu-krisha, the head of the technical department at National Authority for Tunnels (NAT), for his great help to provide the available field and technical data. Authors also would like to thank Dr. Ahmed Hassan (Assoc. Prof., Civil Engineering Department at Faculty of Engineering - Minia University), for his valuable comments.

References

Abu-Krisha, A.A.: Numerical modelling of TBM tunnelling in consolidated clay, Ph.D. thesis, Chapter 7, Innsbruck University, Austria (1998)

Bakker, K.J.: Structural design of lining for bored tunnels in soft ground, Heron J. **48**(1) ISSN 0046-7316 (2003)

Blom, C.B.M.: The structural (un)safety of tunnels explained by analytical approach. Heron **48**(1), 17–32 (2003). Delft University of Technology

Duddeck, H., Erdmann, J.: Structural design models for tunnels, Scribd website 2012, Universitat Braunschweig, Germany (1982, 2012)

FHWA (2009), U.S. Department of transportation – Federal Highway administration (U.S.A), Technical manual for design and construction of road tunnels, Civil Elements, Publication No. FHWA-NHI-10-034 (2009)

Gruebl, F.: Modern design aspects of segmental lining, International Seminar on Tunnels and Ground Works. LNEC, Lisbon, 29–30 June (2006)

Gruebl, F.: Segmental ring design (New challenges with high tunnel diameters), World Tunnel Congress 2012, Bangkok, Thailand, Tunnelling and Underground Space for a Global Society (2012)

Hu, X., et al.: An analytical method for internal forces in DOT shield-driven tunnel. Tunn. Undergr. Space Technol. **24**(6), 675–688 (2009)

Hinchberger, S.D., Ahn, T., Lo, K.Y.: Investigation of segmental concrete tunnel liner joint stiffness. In: World Tunnel Congress and 36th General Assembly, Vancouver, Canada, ITA-AITES, 17 and 19 May (2010)

ITA: WG2 on General Approaches to the Design of Tunnels, Guidelines for the design of tunnels second report, International Tunnelling Association, Working Group No. 2, Research (1988)

ITA: WG2 on General Approaches to the Design of Tunnels, Guidelines for the design of shield tunnel lining, International Tunnelling Association, Working Group No. 2, Research (2000)

JSCE (2007), Japanese standard for shield tunnelling, Japan Society of Civil Engineers, 4 edn., Tokyo (2000). (English translation)

Koyama, Y.: Study on the improvement of design method of segments for shield-driven tunnels (in Japan), RTRI Report: Special No. 33. RTRI, p. 114 (2000)
Koyama, Y.: Present status and technology of shield tunneling method in Japan. Tunn. Undergr. Space Technol. **18**(2–3), 145–159 (2003)
Lee, K.M., et al.: An analytical solution for a jointed shield-driven tunnel lining. Int. J. Numer. Anal. Methods Geomech. **25**, 365–390 (2001)
Mahajan, R.: 3 dimensional structural modelling of segmental tunnel lining, Aust. Tunn. Soc. (ATS) J. (2010)
MIDAS: MIDAS GTS NX V3.0.0 (4), MIDAS GTS Manual, MIDAS Information Technology Co. Ltd., Gyeonggido, 463–400, Korea
NAT: National Authority for Tunnels. Greater Cairo Metro Line III, Project Documents, Cairo (2009)
Wood, A.M.: Tunneling: management by design, London and New York, 2000, Edition 2002 (2002)

Lining Induced Stresses for Mechanized Tunneling Along Curved Alignment

Ahmed Marwan[1,2](✉), Abdullah Alsahly[1], Mostafa Z. Abd Elrehim[2], and Günther Meschke[1]

[1] Institute for Structural Mechanics, Ruhr University Bochum, Bochum, Germany
[2] Department of Civil Engineering, Faculty of Engineering, Minia University, Minya, Egypt

Abstract. In tunneling projects, determining the tunnel alignment is one of the early decisions in the planning process. The alignment is chosen according to feasibility studies of the project. Usually, a curvature of the alignment is inevitable, especially in urban tunneling when certain locations such as exit points, locations with sensitive surface structures and difficult ground conditions must be a priori considered. Mechanized tunneling presents an effective and widely used construction method, in particular, in soft ground conditions and in presence of ground water. A realistic simulation of mechanized tunneling process requires a reliable numerical model that accounts for the different construction stages and all relevant interactions; between the surrounding soil medium and the shield machine during the machine advancement and between the soil and tunnel linings via the grouting mortar.

The stresses in tunnel linings are influenced by construction conditions, ground conditions, and shield thrust. Tunnel lining design requires a proper estimate of structural forces in linings. The segmental linings in curved zones during advancement require special attention to account for expected extra stresses and to avoid possible cracking.

To predict the stress evolution in lining during tunneling process, a 3D finite element model (*ekate*) has been developed which is able to simulate the staged mechanized tunneling process along arbitrary curved alignments. In this paper, the structural loading on a curved aligned tunnel lining segments in different construction stages is discussed. The loading on the lining is directly determined from the modelling of all interactions between the installed tunnel linings and the advancing tunnel boring machine with the surrounding soil. The eccentric thrust forces acting on the lining are mechanically considered in the process-oriented simulation model. The aim of this numerical study is to investigate the influence of these complex interactions on the loading and the induced stresses in the tunnel linings.

1 Introduction

Tunneling projects are commonly exemplifying challenging projects. They usually require major financial costs with high risks and uncertainties. Several construction methods have been developed in tunneling industry and the selection of a suitable

S. Agaiby and P. Grasso (eds.), *Engineering Challenges for Sustainable Underground Use*, Sustainable Civil Infrastructures, DOI 10.1007/978-3-319-61636-0_4

method is a key aspect in the planning phase for a cost-effective construction and successful operational efficiency. Moreover, a successful design of tunnel alignment can considerably improve the constructability of tunnel and decrease its impact.

Generally, the design of tunnel alignment is influenced by the geotechnical conditions as well as the site constraints (FHWA 2009) and a curved alignment is inevitable especially in urban tunneling in which curving the tunnel alignment is a counter measure to bypass either sensitive surface structures, special existing buildings, and landmarks or other underground structures and utilities. Accurate steering of the shield machine, particularly in tight curves, raises an extra challenge during construction (Johnson 2014).

In mechanized tunneling, the tunnel lining has to permanently fulfil some basic requirements concerning structural safety, serviceability, and durability for the entire live time of the tunnel. It sustains various loading conditions such as earth pressure, water pressure, self-weight on the stock, surcharge, and thrust forces. The critical loading case shall be considered by the designer for safe and economic optimal design (ITA 2000). Certainly, the types of shield and geotechnical conditions have a great influence on the possible type of lining. For the geometrical design of segmented concrete lining, there exist different types of segments which differ in assembling and construction stages but do not affect its function. The current study focuses on mechanized shield tunneling in soft ground with precast concrete segments. The traditional standard precast segments with a rectangular plan can only be used in straight aligned tunnels, while tapered trapezoidal segments, instead, allow the construction of three-dimensional curves that follow the designed curved path. More detailed illustrations of various segmented lining types with its different joints detailing, waterproofing and erection can be found in (Maidl et al. 2012; Guglielmetti et al. 2007). The analyses of tunnel lining were studied in the literature using a variety of approaches: laboratory physical model, field measurements, empirical/analytical methods and numerical modelling. For the numerical analysis of tunnel lining, various models have been developed and improved over time. The loading on lining can be evaluated through full two/three-dimensional numerical analysis or by subgrade reaction based models. The lining can be modeled by bedded beams elements, continuum two-dimensional shells or continuum three-dimensional solids. Furthermore, it is considered as a continuous domain neglecting the discontinuities between the segments (Medina Rodríguez and Melis Maynar 2001; Kasper and Meschke 2004; Lambrughi et al. 2012; Katebi et al. 2015). In addition, the evaluation of segments interactions and patterns are presented in the literature (Blom et al. 1998; Mashimo and Ishimura 2003; Klappers et al. 2006; Abd Elrehim et al. 2015). More literature on the analysis of tunnel lining behavior and the influence of the lining joint pattern can be found in (Do et al. 2013).

The purpose of this paper is to investigate the structural loading on lining segments in a curved aligned tunnel for different construction stages. In most cases, models tend to ignore the simulation of the eccentric thrust jacks and its influence on the steering of the machine and on the loading on the lining. As a result, the three-dimensional stress distributions are imprecisely predicted which may cause unexpected damages, cracks and a reduction in structural durability. A computational steering algorithm is used in a process-oriented simulation model. Such a model is capable of simulating the advancement of the shield machine along curved alignment considering the eccentric thrust forces by hydraulic jacks. Such analysis provides possibilities for more precise computation of lining loads. The effect of tunnel alignment curvature on the induced stresses in tunnel linings is demonstrated by numerical example.

The remainder of this paper is as follows; first, the development of the used numerical model of shield machine excavation is reviewed. Second, the modeling of shield machine advancement and the used steering algorithm are briefly presented. Then, the stress transformation to cylindrical stress stat and extrapolation of member forces are illustrated. Finally, a case study of the reference project, Wehrhahn-Line (WHL) metro in Düsseldorf, is considered in order to illustrate the model, and finally the findings in this study are summarized in the conclusion.

2 Computational Mechanized Tunnelling Model

In this paper, the employed numerical model has been developed within the framework of the object-oriented finite element code Kratos (Dadvand et al. 2010); an open-source framework for the development of numerical methods for multiphysics simulations. The process oriented finite element model is denoted as "*ekate*" (Enhanced Kratos for Advanced Tunneling Engineering). This model takes into account all relevant components involved in mechanized tunneling i.e. shield machine, hydraulic jacks, lining, soil medium, ground water, annular gap grouting and face support pressure – and their relatively complex interactions. A brief illustration of the main components of the model is presented whereas a detailed description can be found in (Nagel et al. 2009; Nagel et al. 2010; Meschke et al. 2011). However, such a sophisticated model requires considerable effort to generate. To manage such aspect, the model has been incorporated into the so-called Tunneling Information Model (TIM), which is a Building Information Modeling (BIM) based product, for a more automatized process for the generation and execution of FE simulations (Alsahly et al. 2016).

Figure 1 shows the main components involved in mechanized tunneling (left) and their representation in the finite element model (right). The shield machine is modeled as a three-dimensional deformable body representing its main load-bearing components; the shield skin and the bulk-head. Furthermore, shield skin tapering and its weight are considered. The conical shield skin interacts with the excavated soil by means of the surface to surface contact condition. The shield is supported as well on the lining by means of hydraulic jacks. The jacks are modeled by truss elements. In order to realistically simulate shield advancement, particularly in a curved tunnel routes, an automatic steering algorithm is employed in order to control jacks movements (Kasper and Meschke 2004; Alsahly et al. 2016). Then the jacks are retrieved to assemble a new ring. The sequential excavations are repeated by pushing the jacks onto the newly installed rings, removing the excavated soil elements. After segments erection and shield advancement, the resulting annular gap is usually filled with pressurized grouting mortar.

The distinct phases of the soil medium can be modeled within the framework of the theory of porous media accounting for the coupling between the deformation of the solid phase, water pressures, and air pressure. The inelastic material response is modeled by means of the elastoplastic constitutive law, e.g. Drucker-Prager *plasticity*

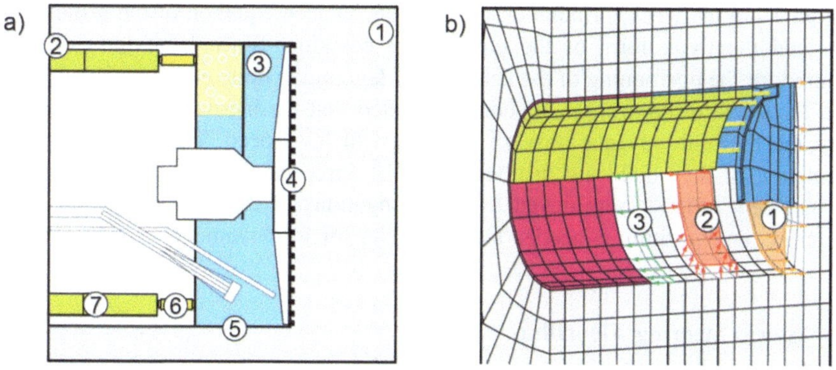

Fig. 1. (a) Main components involved in mechanized tunneling: (1) soil, (2) tail gap, (3) pressurized support medium, (4) cutting wheel, (5) shield skin, (6) hydraulic jacks and (7) segmented lining. (b) Modeling of interactions between soil and TBM in the simulation model ekate: (1) heading face support, (2) frictional contact between shield skin and soil and (3) grouting of the tail gap and components of the simulation model: TBM, hydraulic jacks, lining and grouting mortar

or the nonlinear elastoplastic clay and sand model. The annular gap grouting mortar is modeled by a two-phase material model considering its temporal evolution of the stiffness and the permeability (Kasper and Meschke 2006a; Kasper and Meschke 2006b). The lining is modeled as continuum rings with solid elements and linear elastic constitutive law. Different prescribed boundary conditions at the tunnel face can be accounted for to simulate the various types of face support.

3 Modelling of Advancement Process

Generally, in tunneling projects and especially in machinery tunneling, well-established surveying and positioning systems are a prerequisite to ensure that the tunnel is being constructed according to the designed alignment without exceeding the predefined allowable tolerance. During construction, measuring devices through a set of reference points are used to continuously track the current position and direction of the machine with respect to the designed alignment. Furthermore, positional deviations are instantaneously displayed in graphical and numerical formats on the control screen; thus helps the machine driver to steer the machine and to follow the designed path. The spatial movement of the machine is achieved by jacks' extensions. Such a system, exerting large thrust, is a key part of the shield. It does not only drive the shield but also controls the posture of the shield to ensure that the shield can advance along the expected path consequently for constructing the planned tunnel line (Huayong et al. 2009). In addition, during the advancement of the machine, the tail skin clearance between the segment and machine tail shall be checked regularly.

The motion of the shield machine is governed by the equilibrium of all the acting forces (Festa et al. 2012). In the proposed finite element model, the governing equations are the week formulation of the mass balance equation of the ground water flow and the weak form of the equilibrium equation. Since large movements are required for the positioning of the shield; total lagrangian finite elements formulation is used for shield discretization. It should be noted that the inertial forces are neglected since the machine advance through the soil with low speed. The final position and orientation of the shield are resulted from the force balance on the shield and the advancement process is achieved by the elongations of the hydraulic jacks system. Therefore, an algorithm is needed to control the shield movement.

3.1 Shield Steering Algorithm

The shield machine should be driven as close as possible, with a specific tolerance, to the designed path. In the proposed model, a steering algorithm that controls the shield position has been developed. More illustration of the mathematical formulations used in steering algorithm can be found in (Alsahly et al. 2016). The algorithm provides the instantaneous vertical and horizontal position as well as the driving direction which are used for the subsequent movements. In addition, the non-uniform distribution of the thrust forces can be obtained as simulation results.

Therefore, the resulted thrust forces represent the counter forces acting on the lining to push the shield forward. The computation tunneling simulation model in conjunction with the aforementioned shield steering algorithm enables a more accurate prediction of the time-variant state of stresses on the tunnel lining in particularly for any arbitrary curved aligned tunnels.

4 Induced Stresses in Tunnel Lining

In practice, tunnel lining provides continuous support for the soil medium and prevents water flow inside the tunnel. It is being erected inside the shield while it provides thrust resistance for advancing the shield. Safe and cost efficient tunnel lining designs require a reliable determination of the expected time-variant stresses and deformations in the lining and prognoses of possible critical loading conditions. In this context, tunnel lining has to bear the thrust forces during the advancing of the shield. Particular interest has to be considered in curved parts of tunnel alignment; where eccentric thrust forces can sometimes cause damage in the segments.

As mentioned earlier, the traditional standard segments can be used in straight tunnel and the universal/trapezoidal segments are the most commonly used in both curved and straight alignment, (Fig. 2). Segments geometry should be designed for a specific criterion to follow the designed path as well as to ensure efficient installation. The position of the key segments has to be defined during construction; it could be altered to adjust steering errors (Swartz et al. 2005). A ring tapering is one of the design aspects and it is influenced by the lining width and the radius of curvature.

In the proposed numerical model, the concrete tunnel lining is simulated as continuous rings and discretized by 3D hexahedral elements with the isotropic linear elastic material. It is obvious that the kinematics of the segmented lining is not fully

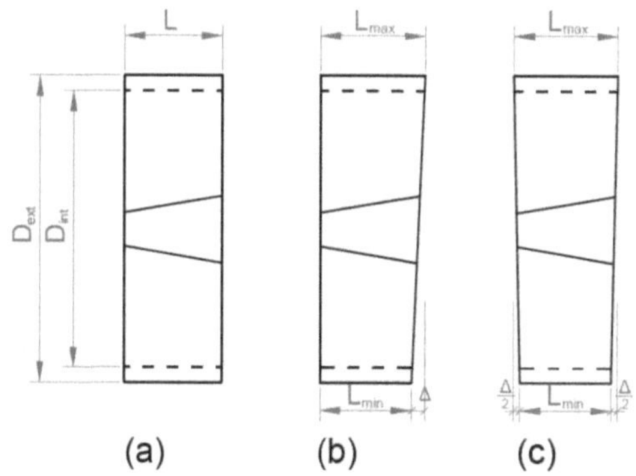

Fig. 2. Types of segmented concrete lining rings (Guglielmetti et al. 2007); (a) straight ring, (b) tapered ring and (c) universal ring

represented by the continuous monolithic cylinder. However, segments interactions are ignored for simplicity. Such choice had been proposed in different kinds of literature (Medina Rodríguez and Melis Maynar 2001; Kasper and Meschke 2004; Lambrughi et al. 2012; Katebi et al. 2015). Such assumption of a uniform rigidity provides the upper bound for structural member forces in particularly bending forces (Koyama 2003; Arnau and Molins 2012). Many kinds of literature studied the effect of joint pattern on segments response; i.e. (Huang et al. 2006; Arnau and Molins 2012; Do et al. 2013). These models tend to substitute the surrounding soil by using spring models which do not represent accurately soil-lining interaction, in addition, dowels and bolts are usually not simulated. According to (Medina and Melis 2002), bolts have an important effect on lining behavior and make it act like a continuum.

In this study, parallel shaped rings are used for the straight part of the alignment, while for the curved part; one-sided tapered shaped rings are generated according to the radius of curvature, Fig. 3. Within the simulation, the evolution of tunnel lining stresses along curved alignment is of interest in order to investigate the influence of the operational thrust forces exerted by shield machine jacks.

To evaluate the tunnel lining stresses in curved alignments, the output stresses are being transformed from the Cartesian coordinate system, defined as $\underline{\boldsymbol{e}} = \left\{\underline{\boldsymbol{e}}_x, \underline{\boldsymbol{e}}_y, \underline{\boldsymbol{e}}_z\right\}$, to cylindrical coordinate system. In straight aligned tunnels, one cylindrical coordinate system can be used where the polar axes are defined by the radius and the angular position along the straight path. On the contrary, correlative cylindrical coordinate systems have to be defined for each consecutive ring along curved paths. In this context, each ring has its own cylindrical coordinate system $\underline{\tilde{\boldsymbol{e}}} = \{\underline{\tilde{\boldsymbol{l}}}, \underline{\tilde{\boldsymbol{\theta}}}, \underline{\tilde{\boldsymbol{r}}}\}$; where the polar axes, $\{\underline{\tilde{\boldsymbol{\theta}}}, \underline{\tilde{\boldsymbol{r}}}\}$, are defined along the central width of each ring, (Fig. 4). The cylindrical coordinate representation can be extracted by the following expression

$$\underline{\underline{\tilde{\boldsymbol{\sigma}}}} = \underline{\underline{\boldsymbol{T}}}^T.\underline{\underline{\boldsymbol{\sigma}}}.\underline{\underline{\boldsymbol{T}}};\ where\ \underline{\underline{\boldsymbol{T}}} = \underline{\boldsymbol{e}} \otimes \underline{\tilde{\boldsymbol{e}}}$$

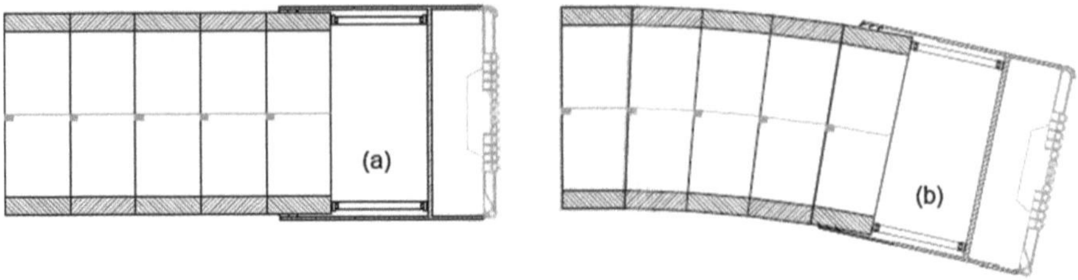

Fig. 3. Sketch representation of tunnel path lining and shield machine; (a) straight path (b) curved path

In practice, the designer uses structural member forces (normal forces, shear forces and bending moments) to check the safety of the lining. The aforementioned cylindrical stress components are used to obtain the structural forces inside each individual ring or a segment.

4.1 Member Forces in Tunnel Lining

In this section, the state of stresses in a specific lining ring is described by means of member forces using the fundamental concepts of structural mechanics. For any arbitrary cross section, the internal member forces in the ring direction, represented by section (a) in Fig. 4, are calculated by

$$N_{\theta\theta} = \sum\nolimits_{elem} \int \sigma_{\theta\theta}\, dA^e, M_{\theta l} = \sum\nolimits_{elem} \int \sigma_{\theta l} * (r - r_{avg}) dA^e$$

where the infinitesimal area vector is defined as

$$dA^e = d\underline{\tilde{r}}\, d\underline{\tilde{l}}$$

Similarly, the internal forces in the longitudinal direction, represented by section (b) in Fig. 4, are calculated by

$$N_{\theta\theta} = \sum\nolimits_{elem} \int \sigma_{\theta\theta}\, dA^e, M_{\theta l} = \sum\nolimits_{elem} \int \sigma_{\theta l} * (r - r_{avg}) dA^e$$

and

$$dA^e = d\underline{\tilde{r}}\, d\underline{\tilde{\theta}}$$

After the calculations of member forces, the most critical sections, according to (ITA 2000), can be nominated as a section with the maximum positive moment, a section of maximum negative moment and section with maximum axial force. The safety of the lining has to be checked in these sections. In addition, the safety against the thrust jacks should be included.

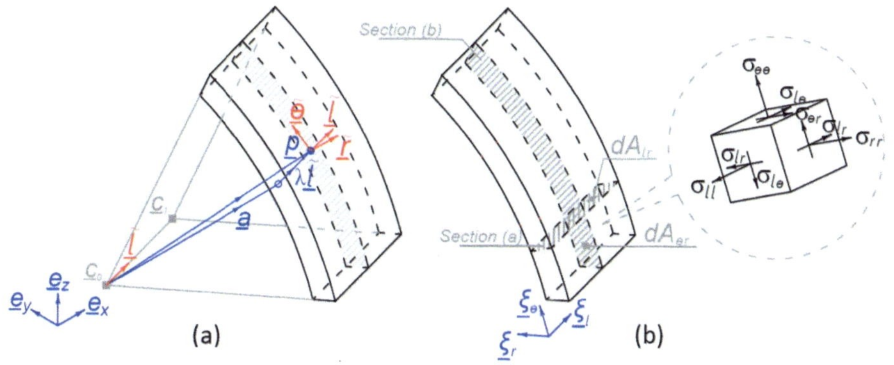

Fig. 4. Induced stresses in lining; (a) cylindrical coordinate system at an arbitrary point and (b) cross sections in ring and longitudinal direction

5 Numerical Tunnelling Simulation

In this section, a finite element analysis of mechanized tunneling along straight and curved paths is performed. The development of the obtained thrust forces are compared for both cases. These forces are automatically extracted from the proposed steering algorithm after each advancing step. Moreover, the time variant stresses and structural forces on tunnel lining are calculated.

The capability of the numerical model is demonstrated by numerical analysis of a reference project, Wehrhahn-Line (WHL) metro in Düsseldorf, Germany. The WHL is a twin-track subway tunnel in the city of Düsseldorf that connects the "Bilk" station in the south to the "Am Wehrhahn" station in the east. The tunnel has a total length of 3.4 km and is divided into a southern branch and an eastern branch, (Blome 2010). A part of the east branch ("Ostast") of the project, more specifically between stations "Schadowstraße" and "Jacobistraße" (Fig. 5a), is investigated in this study. Tunneling was performed by a 9.49 m hydroshield machine under an overburden depth of approximately 1.2 diameter. The tunnel runs through horizontal and vertical curved alignment. In this study, a straight path and horizontal curved path are proposed to demonstrate the capabilities of the model with respect to the three-dimensional steering and the time variant loading on the lining.

For the numerical model (Fig. 5b), the geotechnical investigation is used to characterize soil stratification and the corresponding mechanical and hydrological parameters, Table 1. Drucker-Prager elastoplastic model is employed for the description of the soil behavior.

A linear elastic material model is used for modeling both the lining and the shield machine. The twin-track tunnel has a 45 cm thick 6 + 1 reinforced concrete segments. However, in this paper, lining joints will be disregarded and lining will be modeled as continuous rings. Process related parameters such as face support pressure, grouting pressure and the advancement speed are obtained from monitoring data of the shield during construction. These readings, for the intended section under investigation, are averaged and homogenized over the total simulation domain. The face pressure is

(a)

(b)

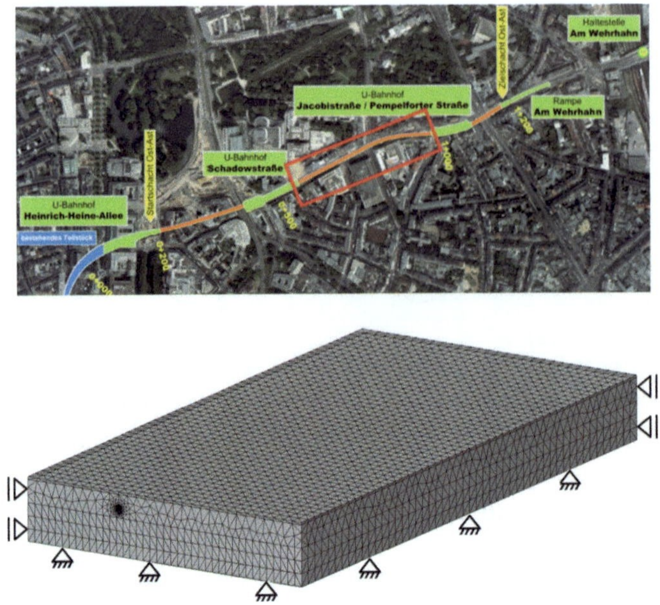

Fig. 5. Wehrhahn-Line in Düsseldorf; (a) route of eastern branch and (b) 3D FE model with displacement boundary conditions

prescribed as constant pressure of 240 kPa on excavation face and on shield face and the 1010 t weight of the machine is applied as a distributed pressure. Ring construction and machine advancement for one excavation step are achieved in a time period of 0.5 h each. The shield is steered with twenty-eight hydraulic jacks. The distribution of the jacks behind the shield is illustrated in Fig. 6, where the jacks are divided into main four groups. The elongation for each jack is obtained from the steering algorithm while the prescribed elongation is averaged for each group of jacks. The realization of machine advancement is performed in 4 successive steps. After shield advancement, the concrete lining ring and grouting is installed. The grouting mortar is simulated as an elastic fully saturated porous material. The pressurization of the annular gap is simulated by prescribing the total stresses and fluid pressure at the shield tail. The prescribed pressure equals 300 kPa and varies linearly with height with a gradient of 10 kPa/m. The time-dependent Young's modulus and permeability are accounting for grouting hydration. The effect of grouting material in tunneling can be found in (Kasper and Meschke 2006; Kasper and Meschke 2006). More comprehensive description of this material model is provided in (Meschke 1996; Meschke et al. 1996).

5.1 Boring Along Straight and Curved Alignments

The aim of this study is to investigate the influence of the driven tunnel path on the distribution of thrust loading on the lining. For this purpose, WHL-tunnel is analyzed employing the steering algorithm. The section under investigation is constructed along

Table 1. Geotechnical parameters used in the numerical simulation

Soil A		
Unit weight γ [kN/m³]	17.32	
Young modulus E [MPa]	21.00	
Poisson's ratio ν	0.25	
Friction angle φ′ [°]	30.0	
Cohesion c′ [kPa]	2.00	Rhine terrace; quaternary sand and gravel, partly minor components of silt and clay
Hardening modulus H [MPa]	5.83	
Porosity	0.40	
Permeability κ [m/s]	0.01	
Soil B		
Unit weight γ [kN/m³]	20.38	
Young modulus E [MPa]	50.00	
Poisson's ratio ν	0.25	
Friction angle φ′ [°]	35.0	
Cohesion c′ [kPa]	0.01	
Hardening modulus H [MPa]	50.00	
Porosity	0.25	
Permeability κ [m/s]	0.001	
Soil C		
Unit weight γ [kN/m³]	21.40	
Young modulus E [MPa]	67.00	Tertiary; slightly silty and medium sandy to silty fine sand
Poisson's ratio ν	0.30	
Friction angle φ′ [°]	33.0	
Cohesion c′ [kPa]	0.01	
Hardening modulus H [MPa]	67.00	
Porosity	0.30	
Permeability κ [m/s]	10^{-6}	

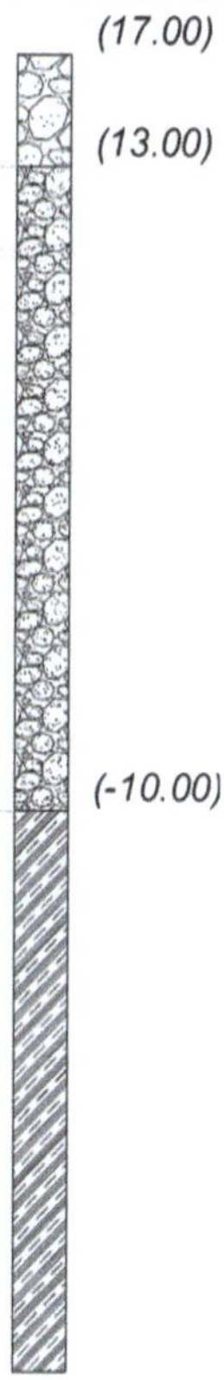

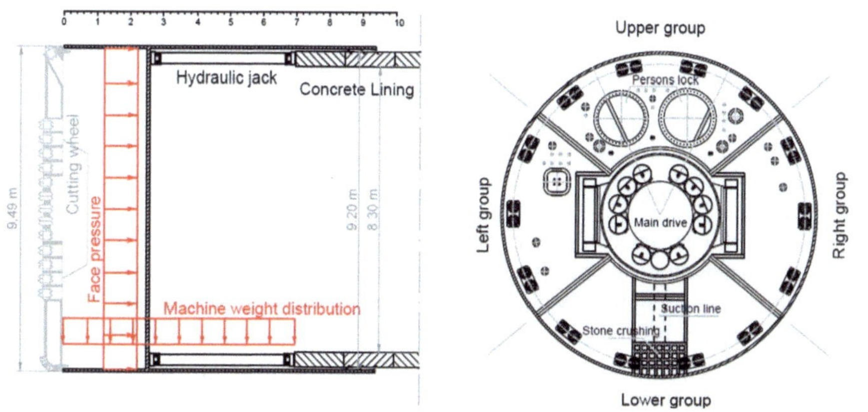

Fig. 6. Wehrhahn-Line in Düsseldorf; sketchmatic draw of the shield machine and the distribution of hydraulic jacks

the straight path then followed by a horizontal rightward curve with a radius of 500 m. The path of the tunnel is defined by a set of points where the steering algorithm uses it to define the target position as well as the required elongation for the jacks. The resulted thrust loading is determined numerically from forces equilibrium on the machine; which represents the counter force acting on the shield (i.e. self weight of the shield, face support pressure and the friction on the shield skin). The output thrust forces distribution, in the straight part of the tunnel, is illustrated in Fig. 7a. In this figure, the exerted thrust loadings are equally distributed. This response is congruous with the distributed face pressure and straightforward advancement. The thrust loading distribution varies with time in order to preserve the equilibrium of the machine. In the curved path, the thrust forces eccentricities are observed in Fig. 7b. The average jack forces for the straight path are 600 kN. While in the curved path, the forces increase to 700 kN for the left group of hydraulic jacks and decrease to 500 kN for the right group. The non-uniform jacks distribution represent the forces induced by the prescribed elongation by steering algorithm. Such a pattern corresponds to advancing along a curve where higher forces are needed to drive the machine to certain position. These forces can be used to check the stresses under the jack shoes according to relevant standards.

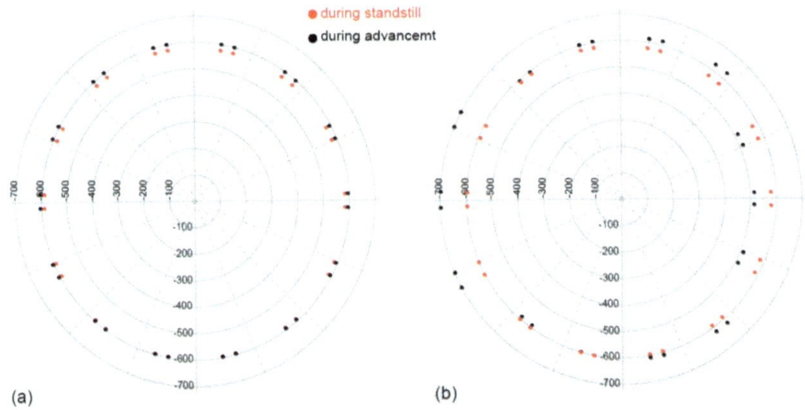

Fig. 7. Thrust forces distribution on hydraulic jacks; (a) straight alignment and (b) curved alignment

Figure 8 shows the computed eccentricities of the total thrust forces from the curved advancement of the machine. The central point cloud corresponds to the stand still position and straight advancement whereas the other point cloud shows bias toward the left side which is consistent with the driven (designed) curved path. The advancement of the shield machine along the designed path is represented in Fig. 9. It can be noted that shield machine settles during standstill, while during advancement, jacks attempt to guide the shield machine into its course accounting for drift off phenomena. In this model, the shield zigzagging movement is captured.

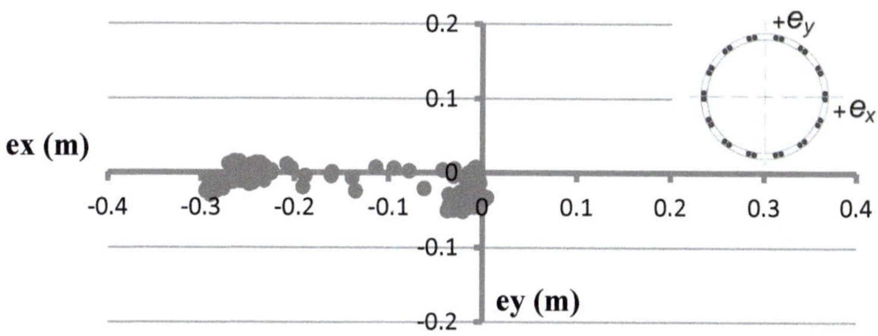

Fig. 8. Computed eccentricities from driving along curved path

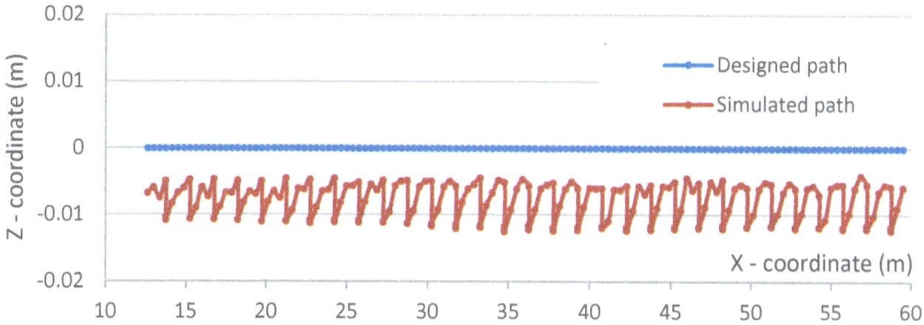

Fig. 9. Numerical advancement of the shield machine (Straight path)

The lining-shield interactions and lining-soil interactions are indicated by the member force acting on the lining. Figures 10, 11, 12 and 13 show the output internal forces in the lining. The normal forces and bending moments illustrated hereafter are with respect to a specific ring on the straight and curved path for both longitudinal and ring directions.

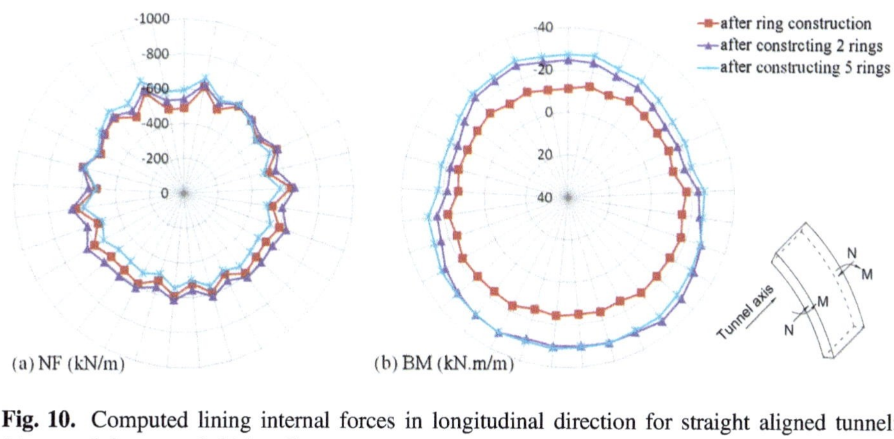

Fig. 10. Computed lining internal forces in longitudinal direction for straight aligned tunnel; (a) normal forces and (b) bending moments

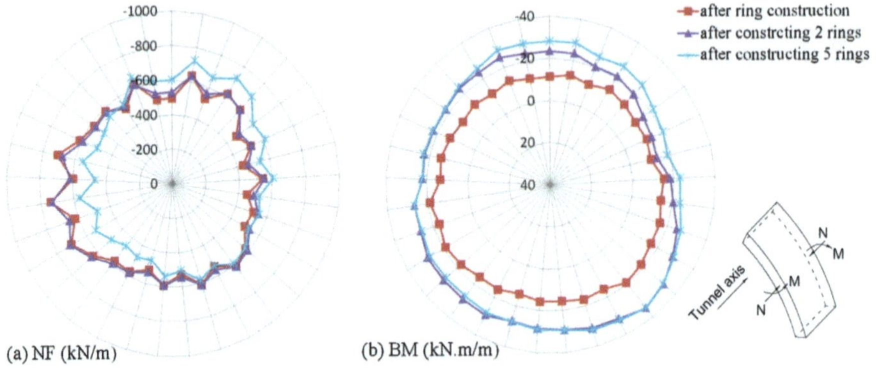

Fig. 11. Computed lining internal forces in longitudinal direction for curved aligned tunnel; (a) normal forces and (b) bending moments

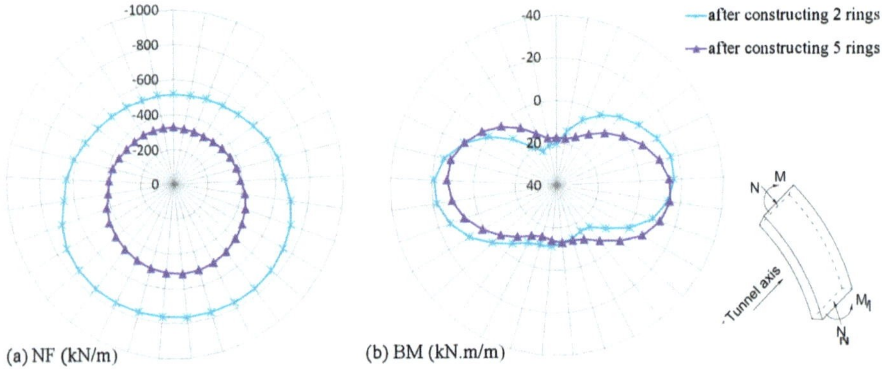

Fig. 12. Computed lining internal forces in ring direction for straight aligned tunnel; (a) normal forces and (b) bending moments

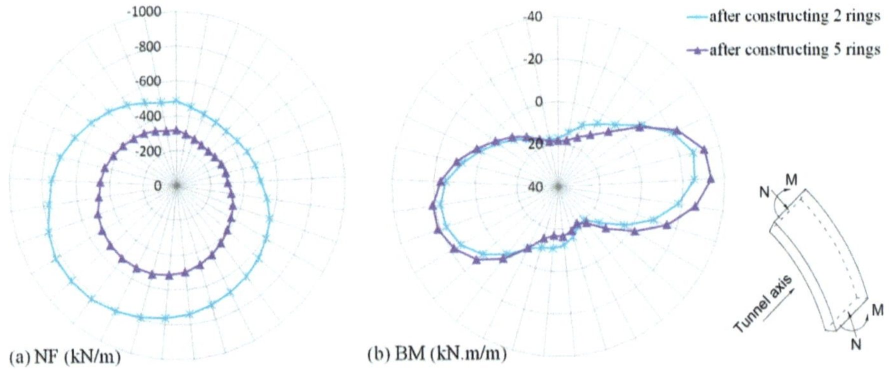

Fig. 13. Computed lining internal forces in ring direction for curved aligned tunnel; (a) normal forces and (b) bending moments

Similarly to eccentric thrust, the output normal forces in longitudinal direction increases on the left side where higher thrust forces. The maximum longitudinal normal force on the currently constructed ring are 637 kN under equally distributed thrust during straight advancement (Fig. 10), it rises by 16% which equals to 738 kN in the curved path (Fig. 11). The resulted moments in the longitudinal direction are not affected with curvature since the hydraulic jacks are simulated in the centerline of the segment. It should be noted that the exact position of the hydraulic jacks with respect to segments centerline has to be accounted for. Comparing Figs. 12 and 13, it can be noted that the ring member forces are influenced by the unsymmetrical loading as well. Figure 12 shows the varying ring forces with respect to time due to grout hydration and pressure dissipation. The highest pressure forces are in the invert of the tunnel since the grout pressure varies linearly with depth. In curved part, the ring forces are concentrated at the lower-left side of the lining, Fig. 13. In addition, the maximum moments and its location change for different driven paths. Such an influence is related to the eccentric thrust loading in the longitudinal direction of the tunnel.

The evolutions of lining longitudinal forces in the curved part are illustrated in Fig. 14. In this figure, the longitudinal normal forces on the left and right sections of a selected ring on the curved path are represented by the olive green and red line respectively. Almost no difference is noticed during the standstill of the machine. While stepped increase/decrease of forces are observed during machine steering. After the ring installation and the following advancement step, the longitudinal normal forces on the left cross-section are higher than these on the right cross-section by 40%. Such a difference is dispersed to 20% after the installation of the third successive ring (time step = 2 h) and dissipated with the installation of the fifth successive ring (time step = 4 h).

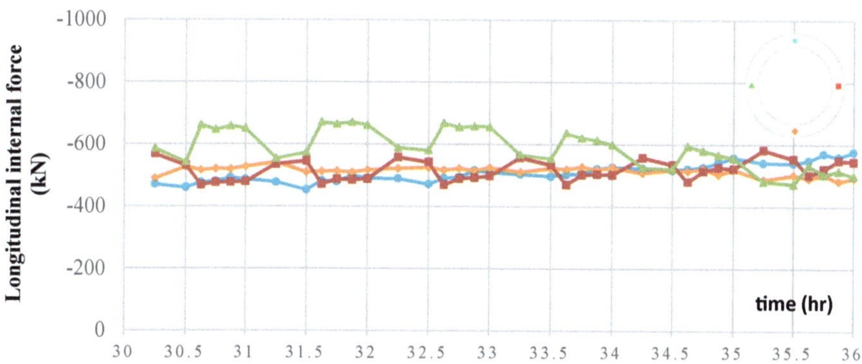

Fig. 14. Evolution of longitudinal forces in a selected ring on the curved path

6 Conclusions

In this paper, the distribution of the thrust forces acting on tunnel lining during arbitrary shield machine advancement in mechanized tunneling process is analyzed numerically. For this purpose, a process oriented finite element model for mechanized tunneling taking into account all relevant components involved in shield tunneling is employed. In this model, the movement of the shield machine during the advancement process is controlled by means of the so-called steering algorithm providing more realistic kinematic of the shield and accordingly more realistic distribution of the thrust forces on the lining during the advancement process considering the soil-shield machine interaction.

For the accurate evaluation of the exerted loads on the lining during the advancement process, the structural member forces are obtained from the 3D stress state and the automatically derived after each advancing step for the individual ring. This leads to more accurate and realistic distribution of the loading which is a prerequisite for efficient and robust lining design. A major advantage of the proposed computational model as compared to existing models is seen in the application of the steering algorithm, which enables the prediction of the thrusting forces during the tunneling process in particular for curved aligned tunnels. The applicability of the proposed model was exemplified by 3D numerical simulation of a straight and curved aligned tunnel and the impact of eccentric thrust is investigated.

From the comparison of the results, it was found, that the distribution of normal forces, in both longitudinal and ring directions, is affected by the distribution of the exerted thrusting forces. In the examined situation for WHL tunnel, the structural member forces and stresses had increased up to 16% in the curved part of the tunnel. In addition, the effect of eccentric thrust forces is dissipated into the ground after the installation of the fifth successive ring. Therefore, it was concluded that the distribution of the acting loads on tunnel lining and segments are affected by the driven tunnel path. A more comprehensive study for various curvatures in horizontal and vertical direction is required to fully quantify such a behavior taking into consideration different friction coefficient between the shield and the surrounding soil as well as the consideration of the process fluid flow around the shield. This may provide valuable information for tunneling engineers during the design phase.

Acknowledgments. Financial support for this work was provided by the German Science Foundation (DFG) in the framework of sub-projects C1 of the Collaborative Research Center SFB 837 "Interaction modeling in mechanized tunneling". This support is gratefully acknowledged.

References

Abd Elrehim, M.Z., Hassan, A., Asaad, M.: Numerical modelling of TBM segmental lining at accidental cases. J. Am. Sci. 2015 **11**(5) (2015)

Alsahly, A., Gall, V., Marwan, A., Vonthron, A.J.N., König, M., et al.: From building information modeling to real time simulation in mechanized tunnelling: an integrated approach applied to the Wehrhahn-line Düsseldorf (2016)

Alsahly, A., Stascheit, J., Meschke, G.: Advanced finite element modeling of excavation and advancement processes in mechanized tunneling. Adv. Eng. Softw. (2016). doi:10.1016/j.advengsoft.2016.07.011

Arnau, O., Molins, C.: Three dimensional structural response of segmental tunnel linings. Eng. Struct. (2012). doi:10.1016/j.engstruct.2012.06.001

Barták, J., Hrdina, I., Romancov, G., Zalámal, J.: Underground Space – The 4th Dimension of Metropolises (2007). doi:10.1201/noe0415408073

Blom, C.B.M., Duurland, H.C.W., Oosterhout, G.P.C., Jovanovic, P.S.: Three-dimensional structural analyses and design of segmented tunnel lining at construction stage. In: Cividini, A. (ed.) Application of Numerical Methods to Geotechnical Problems. ICMS, vol. 397, pp. 87–96. Springer, Vienna (1998). doi:10.1007/978-3-7091-2512-0_8

Blome, A.: Die Wehrhahn-Linie – Historie, verkehrliches Konzept und künstlerische Gestaltung. Vergangenheit trifft Zukunft – 50 Jahre STUVA (2010)

Dadvand, P., Rossi, R., Oñate, E.: An object-oriented environment for developing finite element codes for multi-disciplinary applications. Arch. Comput. Methods Eng. (2010). doi:10.1007/s11831-010-9045-2

Do, N.-A., Dias, D., Oreste, P., Djeran-Maigre, I.: Three-dimensional numerical simulation for mechanized tunnelling in soft ground: the influence of the joint pattern. Acta Geotech. (2013). doi:10.1007/s11440-013-0279-7

Festa, D., Broere, W., Bosch, J.W.: An investigation into the forces acting on a TBM during driving – Mining the TBM logged data. Tunnel. Undergr. Space Technol. (2012). doi:10.1016/j.tust.2012.06.006

FHWA: Technical manual for design and construction of road tunnels — civil elements (2009). Publication No. FHWA-NHI-10-034

Guglielmetti, V., Grasso, P., Mahtab, A., Xu, S.: Mechanized Tunnelling in Urban Areas: Design Methodology and Construction Control. Talyor & Francis, Abingdon (2007). ISBN 978-0-415-42010-5

Huang, Z., Zhu, W., Liang, J., Lin, J., Jia, R.: Three-dimensional numerical modelling of shield tunnel lining. Tunn. Undergr. Space Technol. (2006). doi:10.1016/j.tust.2005.12.076

Huayong, Y., Hu, S., Guofang, G., Guoliang, H.: Electro-hydraulic proportional control of thrust system for shield tunneling machine. Autom. Constr. (2009). doi:10.1016/j.autcon.2009.04.005

ITA: Guidelines for the design of shield tunnel lining. Tunn. Undergr. Space Technol. (2000). doi:10.1016/s0886-7798(00)00058-4

Johnson, N.: Urban EPB tunneling in limited space: a case study of the San Francisco central subway project. In: Proceedings of the North American Tunneling Conference 2014. Society for Mining, Metallurgy and Exploration (2014). ISBN 978-0-87335-400-4

Kasper, T., Meschke, G.: A 3D finite element simulation model for TBM tunnelling in soft ground. Int. J. Numer. Anal. Methods Geomech. (2004). doi:10.1002/nag.395

Kasper, T., Meschke, G.: A numerical study of the effect of soil and grout material properties and cover depth in shield tunnelling. Comput. Geotech. (2006a). doi:10.1016/j.compgeo.2006.04.004

Kasper, T., Meschke, G.: On the influence of face pressure, grouting pressure and TBM design in soft ground tunnelling. Tunn. Undergr. Space Technol. (2006b). doi:10.1016/j.tust.2005.06.006

Katebi, H., Rezaei, A.H., Hajialilue-Bonab, M., Tarifard, A.: Assessment the influence of ground stratification, tunnel and surface buildings specifications on shield tunnel lining loads (by FEM). Tunn. Undergr. Space Technol. (2015). doi:10.1016/j.tust.2015.04.004

Klappers, C., Grübl, F., Ostermeier, B.: Structural analyses of segmental lining – coupled beam and spring analyses versus 3D-FEM calculations with shell elements. Tunn. Undergr. Space Technol. (2006). doi:10.1016/j.tust

Koyama, Y.: Present status and technology of shield tunneling method in Japan. Tunn. Undergr. Space Technol. (2003). doi:10.1016/s0886-7798(03)00040-3

Lambrughi, A., Medina Rodríguez, L., Castellanza, R.: Development and validation of a 3D numerical model for TBM–EPB mechanised excavations. Comput. Geotech. **40**, 97–113 (2012). doi:10.1016/j.compgeo.2011.10.004

Maidl, B., Herrenknecht, M., Maidl, U., Wehrmeyer, G.: Mechanised Shield Tunnelling. Ernst & Sohn, Berlin (2012). ISBN 978-3-433-02995-4

Mashimo, H., Ishimura, T.: Evaluation of the load on shield tunnel lining in gravel. Tunn. Undergr. Space Technol. (2003). doi:10.1016/s0886-7798(03)00032-4

Medina Rodríguez, L., Melis Maynar, M.: Numerical analysis of lining ring behaviour during shield tunnelling. In: FLAC and Numerical Modeling in Geomechanics-2001: Proceedings of the 2nd International FLAC Symposium (2001)

Medina, L., Melis, M.: A numerical analysis of lining behaviour during shield tunnelling. In: Proceedings of the 28th ITA-AITES World Tunnel Congress (2002)

Meschke, G.: Consideration of aging of shotcrete in the context of A 3-D viscoplastic material model. Int. J. Numer. Methods Eng. (1996). doi:10.1002/(SICI)1097-0207(19960930)39:183.3.CO;2-I

Meschke, G., Kropik, C., Mang, H.: Numerical analyses of tunnel linings by means of a viscoplastic material model for shotcrete. Int. J. Numer. Methods Eng. (1996). doi:10.1002/(SICI)1097-0207(19960930)39:183.0.CO;2-M

Meschke, G., Nagel, F., Stascheit, J.: Computational simulation of mechanized tunneling as part of an integrated decision support platform. Int. J. Geomech. (2011). doi:10.1061/(ASCE)gm.1943-5622.0000044

Nagel, F., Stascheit, J., Meschke, G.: Process-oriented numerical simulation of shield-supported tunnelling in soft soils. Geomech. Tunn. (2010). doi:10.1002/geot.201000024

Nagel, F., Stascheit, J., Meschke, G., Gens, A., Rodic, T.: Process-oriented numerical simulation of mechanised tunnelling. In: Beer, G. (ed.) Technology Innovations in Underground Construction. Taylor and Francis, Abingdon (2009)

Swartz, S., Tzobery, S., Himphill, G., Shamma, J.: Trapezoidal tapered ring—key position selection in curved tunnels. In: Proceeding of the Rapid Excavation and Tunneling Conference 2005. Society for Mining, Metallurgy and Exploration (2005). ISBN 978-0-87335-504-5

Tunnel Boring Machine Advance Ground Investigation in Rockburst-Prone Ground Conditions on Neelum Jhelum Project

Irfan Ullah[1(✉)], Gary Peach[2], and Muhammad Nadeem[3]

[1] NESPAK, Lahore, Pakistan
[2] Multiconsult, Oslo, Norway
[3] NDC, Fateh Jang, Pakistan

Abstract. This paper details the methods by which various advance ground investigation methodologies were deployed on two Tunnel Boring Machines (TBMs) working on the Neelum Jhelum Hydro-power project. The project is located in the Azad Kashmir region of northeast Pakistan, in the foothills of the Himalayan range, with overburdens up to 1,870 m and high horizontal stresses, and with complex and challenging geology. The lack of previous project experience in similar conditions, required the use of a wide variety of geological prediction methods and systems. The geological phenomenon of greatest concern was rockbursts: although expected they nevertheless presented a significant danger to personnel and equipment. The ability to predict the likelihood, location, severity and number of rockburst directly impacts upon the tunnel safety and daily production rates. The advanced prediction methods used on this project include forward probing; Tunnel Seismic Tomography (TST), TBM boring parameter analysis, in-situ stress measurements, short term rockburst prediction and continuous microseismic monitoring. This paper describes the various methodologies and how the individual data was integrated on a day-to-day basis to support TBM operations.

1 Introduction

The Neelum Jhelum hydroelectric project is located in the Muzaffarabad district of Azad Jammu & Kashmir (AJK), in northeastern Pakistan. Geographically, the area consists of rugged terrain between 500 and 3200 m in elevation within the Himalayan foothill zone known as the Sub-Himalayan Range.

The project is a run-of-river type, employing 28.6 km long headrace and 3.6 km long tailrace tunnels to cut off a major loop in the river system, transferring the waters of the Neelum River into the Jhelum River, for a total head gain of 420 m (Fig. 1). The headrace tunnels comprise both twin (69%) and single (31%) tunnels, while the tailrace tunnel consists of a single tunnel. Design capacity of the waterway system is 283 cumecs.

The project will have an installed capacity of 969 MW generated by four Francis-type turbines located in an underground powerhouse.

When construction commenced in 2008, all tunnels were to be excavated using conventional drill & blast techniques. However, it soon became apparent that with the

S. Agaiby and P. Grasso (eds.), *Engineering Challenges for Sustainable Underground Use*, Sustainable Civil Infrastructures, DOI 10.1007/978-3-319-61636-0_5

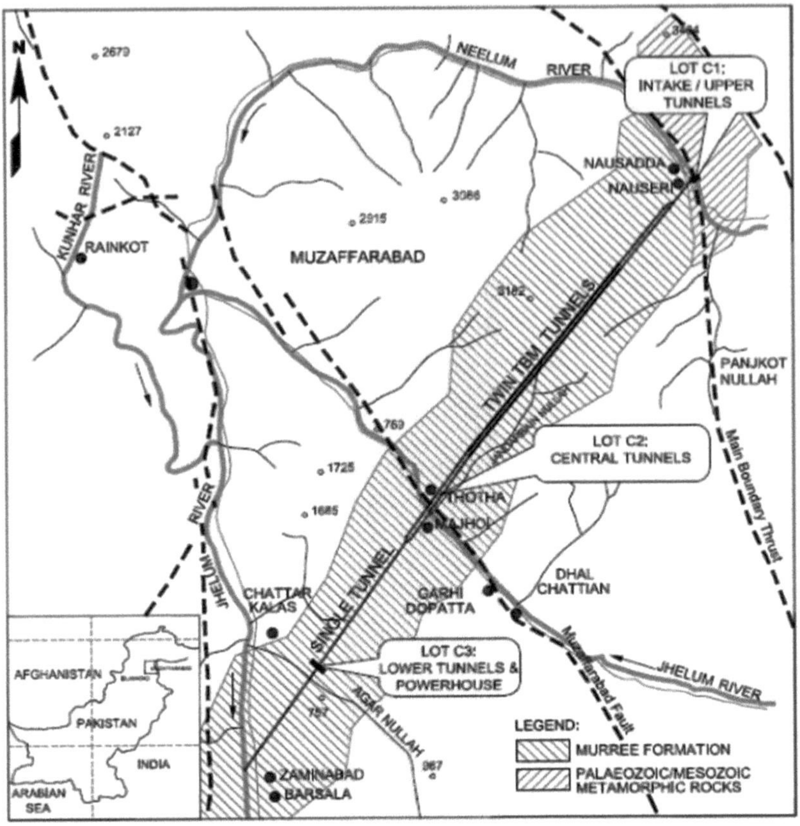

Fig. 1. Neelum Jhelum project layout showing TBM twin tunnels (in bold), major faults (dashed) and alignment geology

equipment being employed, a 13.5 km long section of the headrace twin tunnels underlain by high terrain that precludes construction of additional access adits, would take too long to excavate.

The construction contract was amended to allow the operation of two 8.5 m diameter open (gripper) hard rock TBMs to excavate some 10 km of the twin headrace tunnels (Fig. 1).

2 Geological Setting

The entire project is being excavated in the molasse-type sedimentary rocks of the Murree Formation, which is of Miocene age, and the lateral equivalent of the Siwalik Group in India. At the project site, the Murree Formation consists of intercalated beds of sandstone, siltstone and mudstone that have been tightly folded and tectonized, with generally steep bedding dips and bedding strikes rarely far from perpendicular to the tunnel azimuth. The majority of the rock types, including the mudstones, have a significant calcareous content, typically between 20–30%.

Weakness zones and local faults are commonly observed and are invariably oriented parallel to the regional bedding strike. Sheared zones are generally developed in areas with significant amounts of weak mudstones. These “sheared mudstones”, which often have a well-developed cleavage giving them a shale-like appearance, form a subset of the mudstones encountered throughout the project area and represent the poorest rock types found on the project. While shales are recognized in the literature as a major lithology within the Murree Formation, very few true shales have been identified on the Neelum Jhelum project.

The TBM tunnels are being driven in the central portion of a zone bounded by two major Himalayan faults that trend sub-perpendicular to the tunnels: the Main Boundary Thrust, a bounding fault that extends for the full length of the mountain range that approximately follows the course of the Neelum River at the upstream start of the headrace tunnels at Nauseri (Fig. 1), and a subsidiary reverse/thrust fault named the Muzaffarabad Fault that runs close to the course of the Jhelum River at Thotha. Rupture along this fault in 2005 resulted in the Muzaffarabad earthquake that caused over 75,000 fatalities.

3 Lithologies

3.1 Siltstones and Silty Sandstones

A suite of dark reddish-brown siltstones that grade to silty sandstones constitutes the most commonly occurring rock type on the project, with about 70% of the TBM tunnels excavated in this unit (as of the start of 2016). These rocks typically have uniaxial compressive strength (UCS) values in the 50–70 MPa range.

3.2 Mudstones

Mudstones, which are also invariably dark reddish-brown, represent the weakest rocks in the Murree sequence, with UCS’s in the 30 to 40 MPa range, but with the sheared mudstone sub-group exhibiting significantly weaker properties. Mudstones make up approximately 9% of the total TBM tunnel length, usually occurring as discrete beds of 2 m thickness or less. A review of the geological mapping shows that approximately 16% of the mudstones encountered in the TBM tunnels exhibit shearing, indicating that these poor quality rocks are encountered over only about 1.5% of the total tunnel length, and usually in discrete bands of a few metres’ thickness. Furthermore, careful mapping of a 2.3 km section of TBM tunnel invert that had been left exposed for over a year at the commencement of the drives, showed that only 3.5% of the mudstones (or approximately 20% of all sheared mudstones) showed deleterious slaking tendencies.

3.3 Sandstones (SS-1)

In contrast to the reddish-brown colouration of most of the Murree lithologies, these strongest members of the Murree sequence are usually grey, with sharply defined contacts, both of which make SS-1 sandstones easily distinguishable in exposures and drill holes. Bedding thickness is highly variable, from a few metres to in excess of 50 m. SS-1 sandstone sequences can be continuous along strike over kilometres, and are easily distinguishable in high resolution satellite images as ribs of resistant rock crossing the terrain.

Laboratory test data at the start of the project had indicated mean UCS's of about 90 MPa in this unit, but subsequent tests on samples taken from beds that had produced rockbursts have given significantly higher results. Of the 30 samples tested recently, 77% gave results in the 130–170 MPa range, with the remaining 23% exhibiting higher strengths of up to 230 MPa.

3.4 Groundwater

The SS-1 sandstones form the principal aquifers in the Murree Formation. Nevertheless, despite the fact that approximately 20% of the TBM tunnels has been excavated in this unit, the TBM tunnels have remained mainly dry, indicating limited connection to zones of recharge.

4 In-situ Stresses

In-situ stress measurements for the project were initially directed at the two locations where overburden is most critical, namely at the crossing of the twin headrace tunnels under the Jhelum River and at the underground powerhouse (Fig. 1).

Most of the results were loosely clustered around a horizontal-to-vertical stress ratio (k) of unity, even though high horizontal stresses might have been expected parallel to the tunnel axis, given that the motion of the Indian tectonic plate approximately parallels the tunnel azimuth. Indeed, this is the major principal stress orientation indicated by the World Stress Map (reference 1).

A similar stress regime to that indicated by the ins-situ stress measurements had been anticipated for the TBM. However, subsequent in-situ stress measurements by overcoring in SS-1 sandstone beds in the TBM tunnels, which were initiated to investigate the rockburst phenomenon, have indicated a 'tectonic zone' of high stresses (k up to 2.2) where the major principal stress is oriented sub-horizontally, and nearly normal to the tunnel azimuth.

5 TBM Selection

20 km of the headrace tunnel system was excavated by TBMs in two parallel tunnels. Two open (gripper) TBMs, were used to excavate this tunnel system. The most relevant successful application of TBMs in similar ground conditions was at the Gotthard Base

tunnel in Switzerland. Here, very similar Open (Gripper) type TBMs excavated 85 km of the 157 km rail tunnel system.

At the planning stage, the most significant consideration for the TBM selection was the possibility of encountering squeezing ground, with diametrical deformations of up to 500 mm predicted. This excluded many types of TBM designs because of the risk of entrapment, or the crushing of precast linings. The Open (Gripper) TBM is best suited to deal with this potential condition, due to the short length of the front shield and its ability to collapse inwards various sections of the front shield, depending upon ground conditions, and still maintain the ability to excavate forward.

The second ground condition that required mitigation in the higher overburdens and more brittle rocks was rockbursting. Again, the open (gripper) TBM configuration best allowed for implementation of actions and measures to mitigate against rockbursts.

Furthermore, given the lack of detailed site investigation data because of the high overburden, this TBM design provided the best potential for varying rock support in response to unanticipated conditions.

The designed tunnel excavation diameter is 8.5 m diameter giving a total face area of 56.75 m^2. The tunnel gradient is ascending upstream at 0.8% from SW to NE (Fig. 1).

6 TBM Advance Ground Investigation Equipment

The high overburden over the TBM tunnel alignment precluded cost-effective ground investigations during the feasibility and early construction stages, other than geological mapping of surface outcrops. This mapping was performed for the feasibility study (completed in 1997), and again in more detail at the start of the construction programme. However, given that overburden for most of the route is in excess of a kilometre, prediction of the detailed geology along the tunnel route has simply not been possible, because the tightly folded nature of the bedding prevents reliable extrapolation of geological structures from the surface. This places a greater reliance than usual on advance ground investigations from within the TBM tunnels.

It is preferable that any information on potential adverse ground conditions is obtained as far in advance of the TBM encountering the feature as possible. This allows time to assess the situation and take predetermined counter or contingency measures, allowing the TBM to traverse the adverse condition in the most efficient manner possible, with minimal impact on tunnel progress. The open (gripper) TBMs are flexible to a degree in the range of hard rock conditions that may occur in that, progress is affected when ground conditions that are lower than anticipated competence limits e.g. softer and/fractured rock are encountered. These conditions lead to overbreak in the tunnel, requiring additional and non-standard support as well as cleaning of the debris caused by the overbreak, both of which impact progress. In addition, the rockburst phenomenon represents another ground condition limit that due to its violent and unpredictable nature also impacts TBM progress. These two conditions have come to dominate advance ground investigation systems on this project.

Initially, before the high horizontal stresses that contributed significantly to the severity of the rockbursts were known about, both TBMs were equipped with two separate and independent systems for carrying out advance ground investigation. These were:

- Probe drilling
- Tunnel Seismic Tomography (TST) System

Later in the project when the rockburst zones were encountered, additional ground investigation systems were installed in each TBM tunnel, these were:

- Overcoring for in-situ stress determination
- Microseismic monitoring
- Rockburst prediction based on point load strength determinations

Figure 2, shows the relative overburden along the tunnel alignment (upper portion), and the extent, in plan, of employment of various investigative techniques (lower portion).

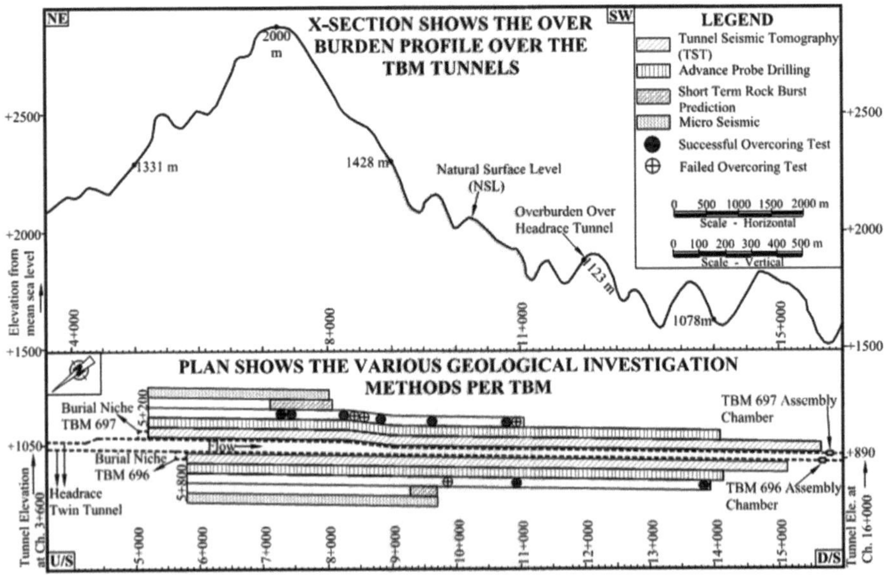

Fig. 2. TBM tunnel alignment and over burden with locations of ground investigations method

6.1 Probing

The design of any probe drilling equipment on a TBM needs to take into considerations the following criteria.

- Flexible and easy to use.
- Effective collaring or starting of probe holes, which is complicated by the shallow entry angles.

- Collaring as close as possible to the tunnel face, avoiding wasteful drilling above already excavated tunnel.
- Ability to move 360° if possible around the tunnel circumference.
- Spare capacity i.e. more than one probe drill to allow simultaneous probing or redundancy in case of breakdown.
- Adequately sized drills with capacity to drill at least 60 m long holes.

Each TBM was equipped with two different types of probe drills, both of which are located at the front of the TBM. One is in a fixed position at the tunnel crown 18 m from the face, and is capable of drilling holes above the shield up 7° from the centerline while the second probe drill is located on a rotary track that enables it to move the full 360° around the tunnel circumference. The probe drilling holes can also be used for any ground improvements such as pre-excavation grouting that may be required.

The principle of probe drilling is that there is always an overlap with the previous hole, so that in no circumstance does the TBM excavate into ground that has not been probed.

The advance information that can be obtained from a probe drilling is:

- Rate of penetration – high indicates weak rock and low indicates hard rock.
- Water inflow.
- Colour of rock chippings – at Neelum Jhelum, grey indicates harder sandstone and red indicates weaker mud rocks.
- Voids and fractured/faulted ground indicated by rapid fluctuations in penetration rate, grabbing of drill rods, and possible loss of flush return.

The information required most critically in the first 2.3 km of TBM tunnels was the presence of softer and or fractured rock that would result in overbreak in the tunnel. During this stage the probe results correlated 80% with the observed geology. After 2.3 km of tunnel excavation, the rockbursts commenced and the prime object became to locate and record the thickness of the SS-1 sandstone beds that are the source of the majority of rockbursts. During this stage the probe results correlated 95% with the observed geology.

7 Tunnel Seismic Tomography (TST) System

The TST system has been developed by Beijing Tongdu Engineering Geophysics Ltd. Corp.

This observation system relies on a series of receivers inserted into predrilled holes on both sides of the TBM tunnel (Fig. 3). A shot is then initiated (Explosive or Electrical) which initiates seismic waves. The reflection of these seismic waves from changes in geological strata are the collected by the receivers. Data processing integrates seismic migration image, velocity scanning and structural grain analysis with travel time inversion imaging, specially designed for tunnel prediction in areas of complex geological structures.

TST is designed to act as a long-range or middle-range forecasting tool, to predict the geological conditions 100–150 m ahead of the TBM. The geological features include, fault structures, interfaces of different type of lithology and rock engineering factors; TST will also highlight the possible areas that may lead to rockbursts.

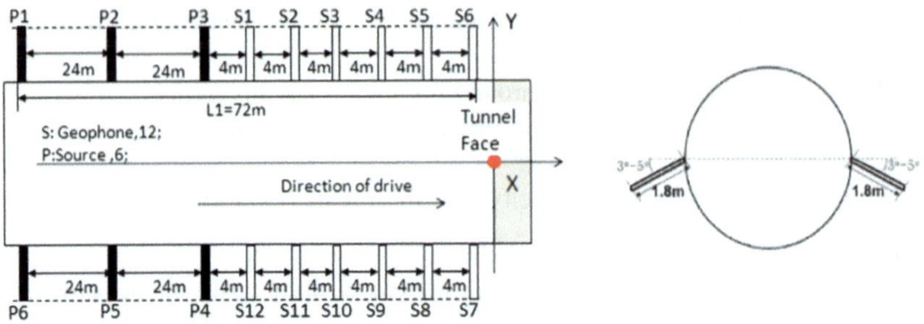

Fig. 3. TST general arrangements for drilling holes (Courtesy Beijing Tongdu Engineering Geophysics Ltd. Corp)

7.1 TST Hardware

TST includes Seismograph, Geophones, Signal Separator, Sparker and data cables (Fig. 4).

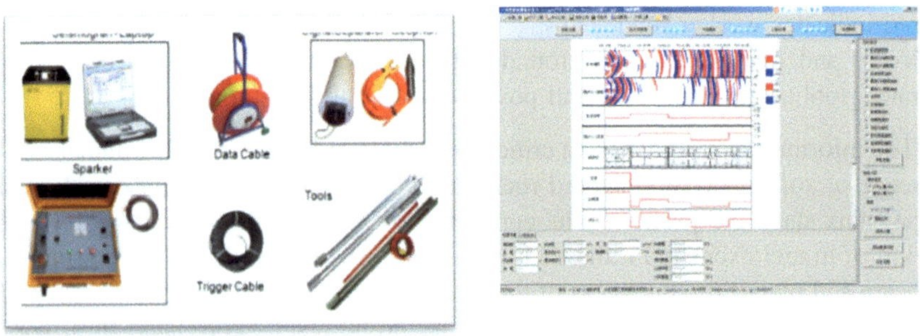

TST Hardware. TST Software Output

Fig. 4. TST hardware and software (Courtesy Beijing Tongdu Engineering Geophysics Ltd. Corp)

7.2 TST Software

TST Data Process Software was developed by the manufacturer (Fig. 4). The system includes: System Setup, Signal Preprocess, Wave Separation, Velocity Analysis, Geological Structure Immigration.

7.3 TST Typical Report

An individual numbered report is produced for each TST operation and should ideally be available within 12 h of the completion of the site operations. The report along with the written interpretation consists of 3 main sections (1) the seismic mitigation image. Here the data is interpreted by adjusting the filtering parameters and an image of significant changes in rock integrity produced. Typically the colours of red and blue are used to denote change boundaries of rock integrity. (2) Velocity through the rock. The higher the velocity the more competent the rock and vice versa. (3) Summary of findings with a prediction of rock integrity with approximate chainages.

7.4 Probe and TST Integration

At the TBM the typical arrangement for implementing a TST survey is every 100–110 m whilst probing is every 45 m (time constraints for production). Both methods have a standard format and the final reports for both techniques are produced within half a day, resulting in both reports being readily available at the TBM for the shift staff to use for day-to-day excavation activities.

In parallel, the site office produces a summary report based on the individual TST report, probe data and observed geology as it is acquired and then carries out a review for any correlation between predicted ground types and observed records recorded as the TBM advances (Fig. 5).

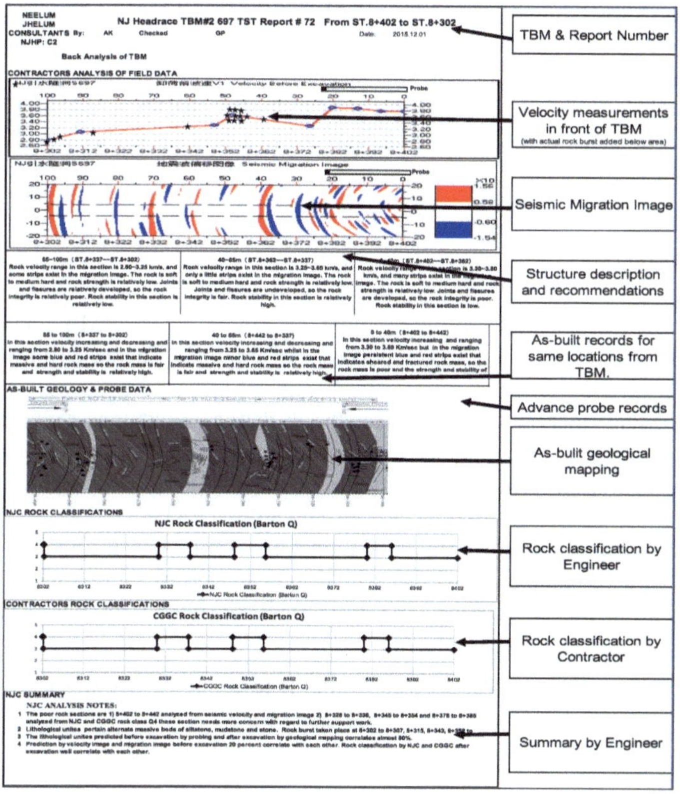

Fig. 5. Typical summary report of TST Table 1. Percentage correlation of TST reports with observed geology

Over the course of the project all TST, probe data and observed geology has been analyzed for the lead TBM for correlation. From the start of TBM tunneling and for the first 2.3 km the significant issue was that of softer and/or fractured ground conditions, from 2.3 km on rockbursts became the significant geological issue. Table 1, gives a summary of the comparison reports.

Table 1. TST correlation with different strata.

Correlation	Average correlation values	Range of correlation values
TST wave velocity and observed geology (soft/fractured ground)	66%	29–72%
TST wave velocity and observed geology (rockburst)	56%	20–80%

During the early sections of the lead TBM tunnel there was correlation between TST wave velocity observed geology, which had resulted in overbreak conditions. This portion of the tunnel did not experience rockbursts and was supported according to the rock class requirements. When rockbursts commenced, the tunnel support was increased and much heavier support was installed. The principle support element being the steel ring beams installed every 1.2–1.6 m with increased shotcrete thickness. In the rockburst section this correlation decreased, the heavier support may well have contributed to this reduction.

8 Rockbursts

No rockbursts were encountered in the initial 2.3 km of either TBM tunnel. After this point rockbursts in both tunnels commenced and gradually increased in number and magnitude.

The Canadian Rockburst Support Handbook defines rockbursts as "damage to an excavation that occurs in a sudden or violent manner and is associated with a seismic event." This definition is widely used in the mining industry, but was considered too specific for the Neelum Jhelum project. Consequently, a rockburst does not necessarily imply that damage has been caused to the tunnel excavation or the TBM: Instead, there may have just been a noise heard that would indicate release of seismic energy. This broader definition was adopted because even small rockbursts can affect the operation of the TBM. The rockbursts have been categorized as follows:

- CAT 1: Noise only: a slight popping sound is heard as energy is released but there no damage to the rock support or ejection of rock

- CAT 2: Noise and weak rockburst: a popping sound is heard as energy is released and there may be some light damage to the rock support and surrounding rock (e.g. cracking in the shotcrete and rock)
- CAT 3: Noise and medium rockburst: loud popping sounds are heard as energy is released and there may be spitting, spalling or shallow slabbing to the rock support and surrounding rock
- CAT 4: Noise and major rockburst: loud sound similar to an explosion, violent ejection of rock into the tunnel and severe damage to the installed support

The summary of the various categories of rockburts are shown in Fig. 6.

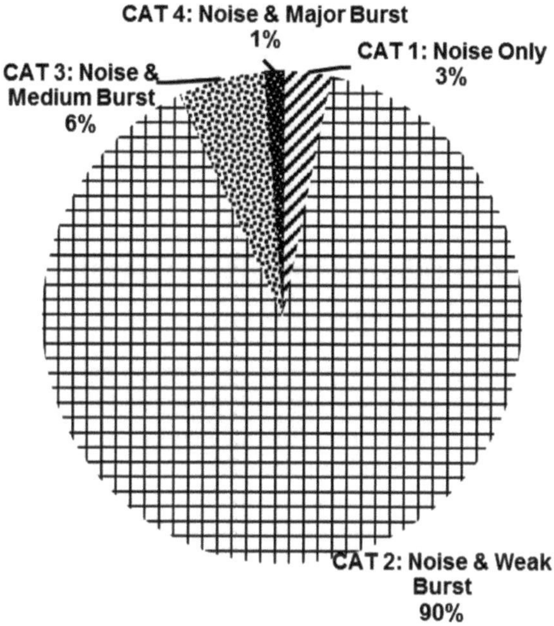

Fig. 6. Chart of Rockbursts by category

Rockbursts, as a phenomenon, were first encountered in the mining industry and so much of the literature relates to rockbursts encountered in deep mining excavations. Traditional mining methods such as drill and blast do not require a high concentration of plant and equipment to be always present near the excavation face. This contrasts with TBM excavation in which complex equipment, with accompanying staff, is continually in close proximity to the excavation face. The TBMs on the Neelum Jhelum project are nearly 200 m long.

When rockbursts commenced, probing and TST results were analysed for location of SS-1 sandstone beds since these were the main source of rockbursts. Additional efforts were then made to analyse the TBM boring parameters as an advance ground investigation tool to identify potential rockburst conditions.

The most severe rockburst on the project occurred on 31st May 2015 on the trailing (by 180 m) TBM. This event, which equated to a 2.4 magnitude earthquake severely damaged the TBM, with the recovery operation taking 7.3 months to return it into operational condition. An unusual set of geological conditions, with the bedding strike locally running parallel to tunnel direction, and unrecognized SS-1 sandstone beds in the sidewalls masked by siltstone, significantly contributed to the severity of this rockburst.

After this event, it was clear that additional ground investigation systems would need to be implemented (and others already ordered would have to be accelerated) to significantly improve the quality of identification of rockburst-prone locations ahead of each TBM. These systems would then allow the TBM crews to implement rockburst countermeasures aimed at lessening the potential rockburst effects, with the added benefit of increasing the crews' confidence by the identification and consequent mitigation of potentially hazardous conditions.

9 Correlation of Rockbursts with TBM Operating Parameters

When rockburst activity became a regular occurrence, an effort was made to identify correlations between boring data and the rock mass conditions encountered, in particular those associated with rockbursts. The findings of the analysis were adopted for guidance in determining the operating parameters of the TBM which would identify strata where the likelihood of rockburst activity was high.

There are five main TBM boring parameters to consider:

- Thrust force (F)
- Torque (T)
- Penetration rate (P)
- Advance speed (V)
- Cutter head Rotation

The TBM data acquisition system measures Thrust Force, Torque, Cutterhead rotation and Advance speed directly from sensors. Whist the Penetration rate is a calculated value obtained by dividing the advance speed by the Cutter head rotation.

Every rockburst time was recorded and then plotted against four of the above TBM boring parameters. The parameters have also been subdivided into bands associated with ground conditions. The results of this analysis are shown in Fig. 7.

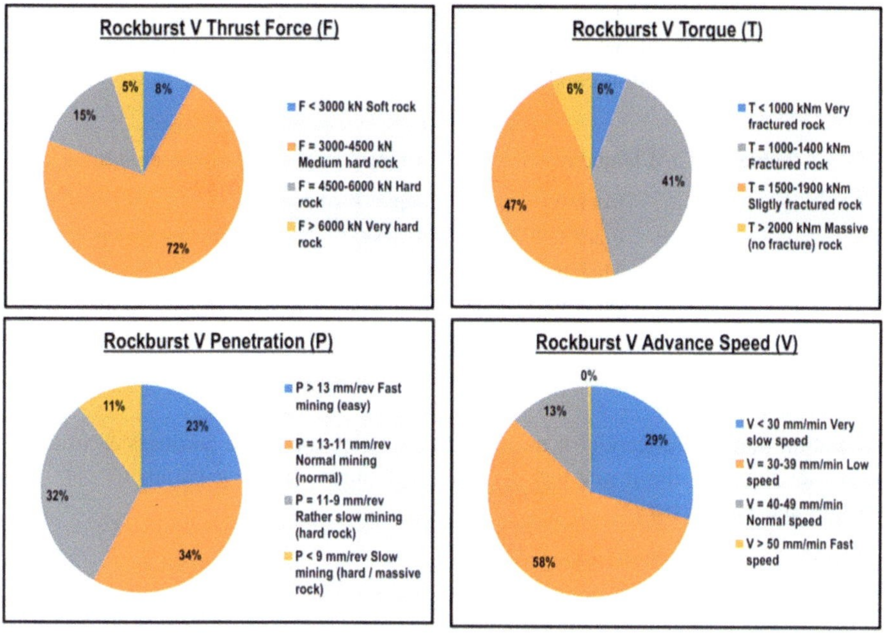

Fig. 7. Rockbursts recorded against TBM parameters.

9.1 Thrust Force

Thrust force is the force applied by the thrust cylinders to the cutterhead to cut the rock when excavation starts. The value of the thrust force depends on the strength (hardness) of rock mass i.e. the harder the rock mass, the greater the thrust force needed to be applied by the TBM operator to excavate it. The average ranges of the thrust force in mudstone, siltstone and sandstone are shown in Table 2.

Table 2. TBM thrust force for different rock types on Neelum Jhelum project

Rock type	Thrust force (kN)
Mudstone	F < 3000
Siltstone	2750 < F < 3750
Sandstone	F > 3500

9.2 Torque

Cutterhead torque is the power from the cutterhead motors that rotate the cutterhead in order to crush and cut the rock mass at the cutterhead. The torque value depends on the structure and roughness of rock mass. If the rock mass is soft, fractured and/or the surface smooth the torque value will be low. However, if rock mass is hard, massive, and/or surface rough, the torque will be high, because of the resistance of the rock mass to the cutter discs arranged around the cutterhead. Furthermore, if the rock in front of the TBM cutterhead collapses, torque value also will increase. The average ranges of torque in mudstone, siltstone and sandstone are shown in Table 3.

Table 3. TBM torque for different rock types on Neelum Jhelum project

Rock type	Torque (kNm)
Fractured	T < 1000
Medium	1000 < T < 1500
Massive & rough	T > 1500

9.3 Penetration

Penetration rate is a reflection of the condition of rock mass, the hardness & structure, the harder the rock the lower the penetration if other operating parameters are kept constant and vice versa.

9.4 Advance Speed

If the advance speed is too high in weaker ground then the spoil conveyance system e.g. Conveyor belt can become over loaded or even blocked. The support installation can also impact advance speed since it is not good practice to advance fast and then stop abruptly whilst support is installed. Each project develops its operational boundaries. This parameter is directly controlled by the TBM operator in accordance to the condition of rock mass, machinery, and transporter of spoil material.

9.5 Cutterhead Rotational Speed

Rotational speed (or speed of rotation) is defined as the number of revolutions the TBM cutterhead completes per minute. The rotational speed directly influences the size of the excavated tunnel material. The rotational speed should be controlled in accordance to the rock mass criteria, i.e. rock hardness and rock structure. Rotational speed should be increased if the rock mass is getting harder, and should be decreased in more fractured rock mass. This is important to avoid any damage to the cutter discs and/or damaging the tunnel perimeter by plucking. If the rock mass becomes harder, the TBM also needs to increase the thrust force to achieve required penetration, as explained above. This also leads unavoidably to more pressure on the TBM cutter discs. To try and balance thrust force and torque, rotational speed should be increased. If this is not done, then the TBM cutter discs could possibly block up and stop turning, resulting in damage to the cutter disc by becoming flat. In a softer rock mass or fractured ground, rotational speed should be decreased since too high a speed will probably result in spalling or overbreak.

The relationship between rotational speed, thrust force and torque are shown in Tables 4 and 5.

The time of the rockburst was recorded and compared against the particular TBM operating parameter. All the above data was then plotted against the observed geology and analysed for correlation with the two significant geological issues of (a) soft/fractured rock and (b) rockbursts.

Table 4. TBM thrust force and rotational speed on Neelum Jhelum project

Thrust force (kN)	Rotation speed (rpm)
F < 3000	rpm < 3.0
3000 < F < 4000	3.0 < rpm < 3.5
4000 < F < 5000	3.5 < rpm < 4.5
F > 5000	rpm > 4.5

Table 5. TBM torque and rotational speed on Neelum Jhelum project

Torque (kNm)	Rotation speed (rpm)
T < 1000	rpm < 3.0
1000 < T < 1500	3.0 < rpm < 3.5
T > 1500	rpm > 3.5

9.5.1 Summary

The conclusion of the analysis and examination of all boring parameters at the locations where rockbursts occurred indicate the way TBM is operated appears to have little to do with influencing rockbursts. A rockburst will happen when the rockburst conditions are met, i.e. (1) amount of energy accumulated around the perimeter is sufficient, (2) direction of energy, towards weak zone, and (3) line or zone of weakness to release energy and tangential stresses exceed the compressive strength of the rock mass. If all three of these factors reach a certain threshold, then a rockburst will happen, whatever the TBM operating parameters; it will make little difference if the TBM is advancing fast or slowly.

However, by analyzing the boring data there are some trends which can be used to indicate the likelihood of a rockburst occurring.

- Rockbursts occur in very hard and brittle rock masses that also have few fractures (i.e. massive). These parameters correspond to thrust forces from 3000 kN to 6000 kN, and torques from 1000 kNm to 1900 kNm. Recognition of these parameter ranges alerted TBM operators to the presence of rock masses likely to cause rockbursts. This is more important than may first appear, because while probe drilling identified the presence of an SS-1 sandstone bed, it rarely gave an indication of the degree of fracturing. In contrast, the TBM cutterhead by the very nature of its design and operation, provided instant identification of the rock mass quality of the ground at the cutterhead that could be directly correlated with its rockburst potential.

- Rockbursts will not happen in soft rock masses or a highly fractured rock mass, i.e. those that result in thrust forces below 2750 kN, and torque below 750 kNm.
- Counter intuitively, rockbursts do not appear to occur in very hardest and most massive rock masses, where the applied thrust force and torque are above 7500 kN and 2500 kNm respectively. This may be because the rock mass is simply strong enough to withstand the imposed tangential stresses.

A note of caution needs to be added on the plotting of rockbursts against the TBM boring parameters data taken from the TBM data acquisition computer. The data shown is the average data throughout the TBM operation for a single stroke (one advance cycle that typically takes 45 min). In contrast, the rockburst is a sudden event lasting milliseconds.

10 In-situ Stress Measurements by Overcoring

The decision had already been made to implement stress measurements by overcoring to gain a better understanding of the rockburst phenomenon when the severe rockburst of 31st May 2015 occurred. This event resulted in the acceleration of the implementation of this technique. The Institute of Rock and Soil Mechanics (IRSM) from the Chinese Academy of Sciences was contracted to perform the overcoring tests in the TBM tunnels.

Ten locations were initially proposed for overcoring tests, concentrating on the SS-1 sandstone beds where a major rockburst had occurred. For this purpose eight locations were selected from the leading TBM and two from the trailing TBM. At eight of these locations, the rockbursts had resulted in damage of different categories to the TBM and tunnel installed support, and the remaining two locations were selected on the basis of overburden depth.

11 Number of Tests at Each Location

After discussion with IRSM, it was decided to perform three tests in a single borehole at each location, following IRSM's standard operating procedure. The minimum depth specified for the first test was 17 m, which is twice the TBM tunnel diameter. Each overcoring test required a 0.5 m long section of borehole.

12 Instrument Used for Stress Measurement

The Hollow Inclusion cell (HI cell) was used for the overcoring tests. This instrument measures micro strain using 12 strain gauges during and after the overcoring test. The data acquired is recorded on a laptop located at the test location. The elastic properties of the rock are measured using a biaxial test on the overcore specimen. During the biaxial test, the applied stresses are increased from 0 MPa to 20 MPa at intervals of 5 MPa and then unloaded to 0 MPa gradually. From this data, the in-situ stresses can be calculated by applying Hooke's law.

12.1 Core Disking

Core disking was observed during the drilling of a number of bore holes in sandstone in both TBMs under different overburdens in Fig. 2 the overcore failed tests were due to core disking. It is a process in which drilled core breaks into disks with uniform spacing and shape (Fig. 8). Core disking is indicative of high stresses in the borehole that were released during drilling of core. The bore holes in which core disking was observed have higher stresses compared to the bore hole where core disking was not observed and the overcoring test was performed successfully.

Fig. 8. Core samples showing disking

Clearly, measuring the overcoring technique relies on the core staying completely intact. The presence of a single discontinuity, even if it is healed, can disqualify a test. Consequently, the breaking up of drill core into small disks totally precludes any overcoring tests. In the lead TBM, four overcoring tests were terminated due to core disking in bore holes. Whereas in the trailing TBM, one bore hole was terminated due to core disking. In all these bore holes, the core disking commences from 10 m to 14 m and continued to the end of the hole.

12.2 Summary

The overcoring results shows that the maximum principal stresses are sub-horizontal, and act from northwest direction, which is perpendicular to the tunnel alignment. These are the tectonic stresses. These stresses range from 73.9 MPa to 141.2 MPa. The intermediate principal stresses are parallel to the tunnel alignment.

13 Short Term Rockburst Prediction (Uniaxial Compressive Strength)

The aim of short term rockburst prediction was to attempt to link various geotechnical parameters identified in the already excavated tunnel immediately behind the TBM shield with other parameters. IRSM staff recorded various geotechnical parameters and after analysis issued rockburst prediction reports on a daily basis for both TBMs. The rockburst prediction reports covered 12 m of tunnel section, 6 m excavated section over the shield of TBM and 6 m unexcavated section in front of TBM cutterhead.

13.1 Parameters for Rockburst Prediction Analysis

The following geotechnical parameters of the excavated ground were used in the short term rockburst prediction analysis.

- Lithology.
- Uniaxial compressive strength (UCS).
- Geological Strength Index (GSI).
- TBM boring parameters (Cutterhead thrust, Penetration and Torque).
- Rock density.
- Uniaxial tensile strength of rock units (Laboratory test).

On the basis of numerical calculation and analysis of geotechnical parameters, the rockburst prediction was categorized into slight, medium and severe rockburst.

13.2 Effectiveness of the Short Term Prediction Technique

The short term rockburst prediction reports were produced and discussed at the daily meetings over the period January – April 2016. During this period many attempts were made to integrate the results with other forms of advance ground investigation and produce a concise rockburst prediction for both TBMs. The results when compared to other investigation methods such as the continuous microseismic monitoring, did not enhance the information already being supplied. Therefore this advance ground investigation method was stopped.

14 Microseismic Monitoring

The project team was concerned about placing too much reliance on one system. Consequently, it was decided to investigate prediction techniques that have an established track record, and microseismic monitoring was such a technique. It was anticipated that using the short term rockburst technique and microseismic technique in parallel would obtain the best possible result in predicting rockbursts.

IRSM, the institution that provided the staff and equipment for the stress measurement overcoring, and for the trial of the short term rockburst prediction, also specialized in providing microseismic monitoring services. IRSM employed the same technique for the Jinping II project which utilized similar TBMs and experienced extensive rockbursts. Several technical papers on microseismic monitoring (Feng et al. 2012, 2015) were researched and contact was made with IRSM to install similar monitoring systems on the two Neelum Jhelum TBMs (Fig. 9).

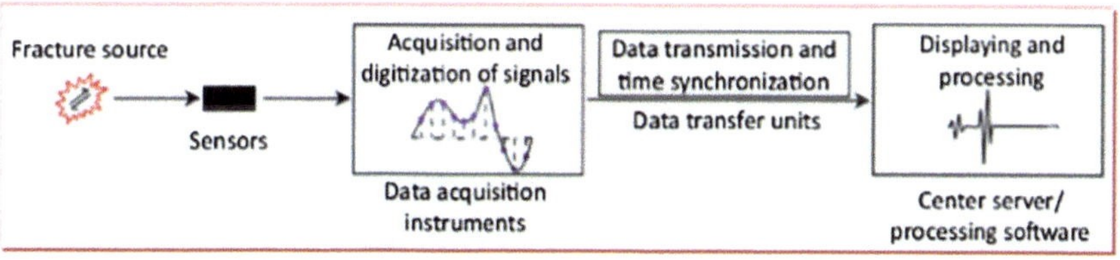

Fig. 9. Microseismic monitoring equipment installed on NJ Project (Courtesy of IRSM)

This paper does not go into detail concerning the development of microseismic monitoring, which is adequately covered by various technical papers by Professor Xia-Ting Feng. The principle of microseismic monitoring is the recoding of the fracturing process in rock masses caused by microseismicity in the surrounding rock. A series of sensors is positioned in the tunnel behind the TBM. A typical configuration is 3 sensors in an array and 3 arrays located between 8 and 60 m from the tunnel face. Theses arrays are moved forward as the tunnel advances. The sensors record the fracture data and transmit to an acquisition systems which digitizes the data. The digitized data is then transmitted to a central server and processed by IRSM software.

The outputs on a daily basis produced from IRSM's microseismic monitoring are shown in Figs. 10 and 11 along with a written statement on rockburst level and

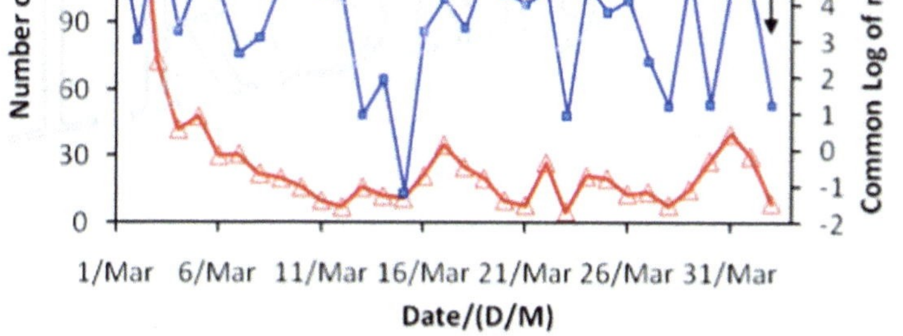

Fig. 10. Number and accumulated energy against time (Courtesy of IRSM)

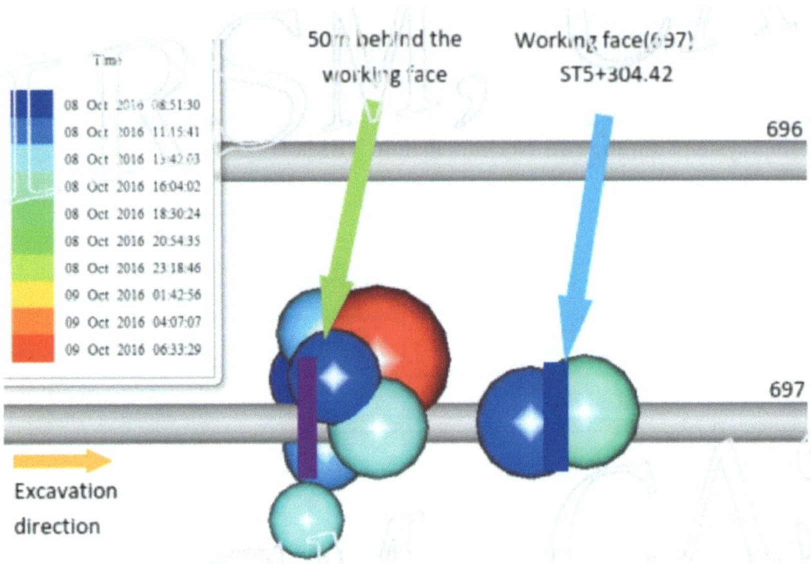

Fig. 11. Temporal-spatial distribution of microseismic activity. Colour and size of sphere indicating time and magnitude of event *(Courtesy of IRSM)*

suggestions for support activities. Figure 10 provides the data on (1) number of seismic events recorded and (2) accumulated energy per event. These two sets of values are produced on a rolling daily basis. This data provided evidence of increasing likelihood of rockburst by increasing numbers of events. Whilst the trend of decreasing number of events recorded and increased average energy per event would indicate the possibility of a larger category rockburst. Figure 11 shows the temporal-spatial distribution of the recorded events. This information is recorded in relation to the TBM working face. Microseismic activity increasing on one the left or right side of the TBM tunnel would help decide priorities for tunnel support or location of counter measures With regard to microseismic monitoring, the authors of this paper aimed to integrate all advance ground investigation methods and expand on proven systems.

14.1 IRSM Monthly Report

In the same manner as the correlation report for TST and probing, all the above microseismic data along with actual rockbursts recorded at the TBM and the observed geology was plotted on a one page summary report. This allowed back analysis and discovery of trends and patterns in microseismic recorded events and conditions observed in the TBM tunnel. A microseismic summary report was produced on a monthly basis (Fig. 12).

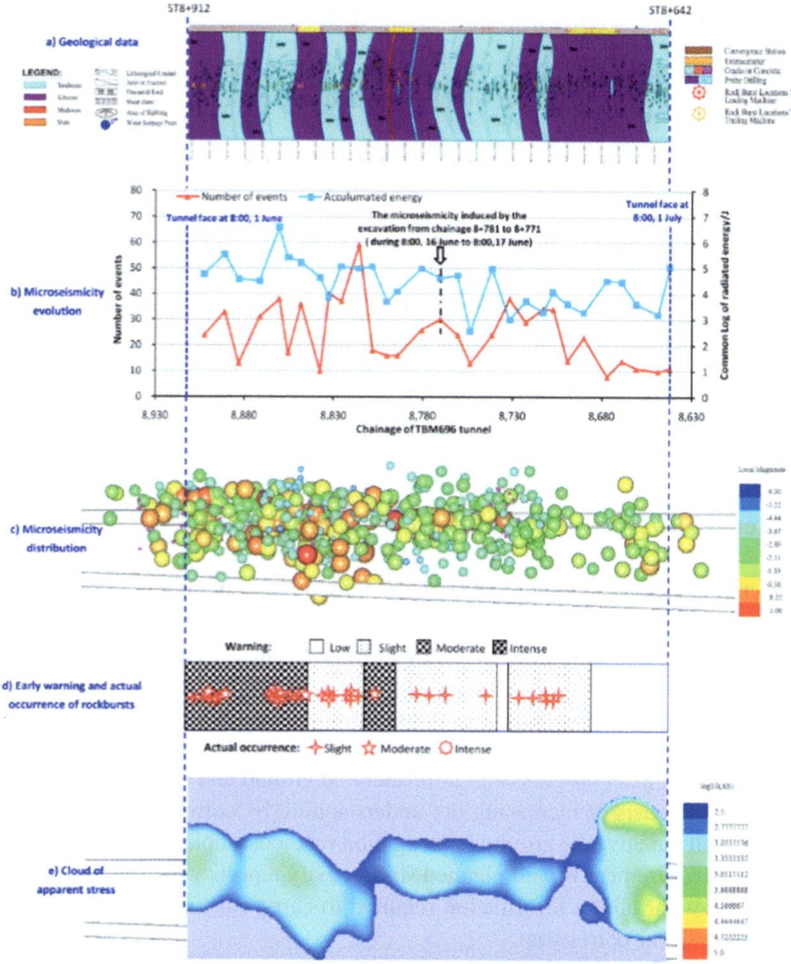

Fig. 12. Summary report for microseismic activity against observed geology for June 2016 (*Courtesy of IRSM*)

A scale mark of 50 m was added to the Temporal-spatial distribution of the microseismic activity chart. This allowed a detailed analysis of how many microseismic events took place in (1) in front of the TBM excavation face (2) from the TBM face back along the tunnel for 50 m and (3) beyond 50 m from the tunnel face.

14.2 Seismic Event Location

Further analysis was carried out to separate the location of the microseismic events into the following locations (1) Open side of TBM tunnel, (2) Above TBM Tunnel and (3) Pillar side of TBM tunnel.

14.3 Microseismic Events and Recorded Rockburst Events

Microseismic monitoring records the full range of seismic events; from those that cannot be heard by the human ear to the severe rock bursts.

14.4 Summary

Microseismic monitoring is an important mechanism for attempting rockburst prediction. However, this science is still in the development stage. The systems employed and the subsequent on-site adaptations have greatly increased the confidence in the prediction probability for a rockburst event. The day to day analysis uncovered more trends and patterns which have allowed both TBMs to advance safely and efficiently in rockburst prone ground conditions. During this stage the microseismic results correlated 88% with the observed geology in which rockbursts occurred.

15 Intergrating All the Data

This paper illustrates the variety of advance ground investigation methods that are available for use in a TBM tunnel to predict the geology in front of a TBM. They include established methods such as probe drilling and newer methods still undergoing development such as microseismic monitoring. The key issue is how to integrate these methods in such a way that they complement each other and contribute to a greater and more accurate understanding of ground conditions and specific geological entities in front of the TBMs that could impact progress e.g. Soft/fractured ground and rockbursts. The project team built on the basis of increasing the understanding by correlation of the observed geology with all the advance ground investigation methods implemented on the project.

The basic requirement is well planned shift/activity reports which provide in a clear and concise manner all the information required to carry out initial data interpretation and then upper level correlation.

Correlation with observed geology remained continuous throughout the project for Probe with TST and TBM boring parameters, the only modification was the change in focus on the correlation with overbreak and rockbursts.

The initial integration involved the probe drilling, TST data with the observed geology. The TST gave an indication of changes in rock integrity that could be checked when a probe was drilled in the same location. The observed geology in the tunnel would then confirm or not confirm any correlations between the two systems.

The TBM boring parameters were plotted against the observed geology. This resulted in the identification of operational bands within which balanced TBM equipment ware and tunnel excavation profile. This information was then available whilst the TBM was excavating.

When rockbursts commenced, it was beneficial that correlation of two investigative techniques had already been established and was in regular operation. This facilitated quick implementation of new geological investigation systems to the adopted formatto identify potential rockburst locations, and thus guide the application of counter or contingency measures.

The overcoring tests to determine the in-situ stresses primarily measured at locations of rockbursts, while not a routinely applied investigative tool, discovered higher than expected horizontal stresses. This information was significant and directly contributed to the application rockburst support.

Subsequent advance ground investigation methods attempted were short term rock burst prediction by point load strength testing, and continuous microseismic monitoring. Both methods commenced at similar times but the continuous microseismic monitoring proved to be the more reliable of the two methods. While the short term rockburst prediction system, which was unable to determine conditions ahead of the face, was dropped due to low than required correlation with rockbursts.

After adoption, the data generated by the microseismic method was correlated with the probe and TST data with the specific aim of identifying rockburst-prone locations ahead of the face. This correlation proved very successful. This led to further analysis to indicate any trends in location along the tunnel and across the tunnel which allowed safety recommendations to be issued concerning certain locations within the TBM.

All the above data was made available at the 08.30 am morning meeting, which allowed discussion and recommendations to be developed before the start of tunnel excavation at 12.30 pm. After the morning meeting the new data would be added into the particular correlation process for analysis, results being fed back into the following day's morning meeting.

16 Conclusions

There is no single advance ground investigation method that can completely predict the ground conditions in front of a TBM. There are however a number of systems that can give an insight into a particular aspect of the geological conditions that lie in front of the TBM. When the results of these methods are integrated, the prediction can be comprehensive.

The key issue was the cross correlation with other methods and observed geology in a standard format which allows the identification of patterns and trends. This may well be unique to this project. Some of the methods used were more successful than others, but various methods have been tried and implemented.

With particular regard to rockburst prediction it is still a developing science and on the project the use of every available advance ground investigation method yielded pertinent information which helps facilitate an increasing ability to predict the probability of a rockburst.

References

World Stress Map Project. http://www.world-stress-map.org

Feng, G., Feng, X., Chen, B., et al.: A microseismic method for dynamic warning of rockburst development processes in tunnels. Rock Mech. Rock Eng. **48**, 2061 (2015)

Feng, X., Yashunb, X., Feng, G.: Mechanism, warning and dynamic control of rockburst evolution process. In: ARMS7 – 7th Asian Rock Mechanics Symposium (2012)

Improving the Existing Roadway Tunnels Capacity by Adding New Tunnels – A Structural Approach

Mostafa Zaki AbdElrehim[1(✉)], Mohamed A. Eid[1], and Osama Moshref[2]

[1] Civil Engineering Department, Minia University, Minya 61111, Egypt
mostafa.zaki@mu.edu.eg
[2] Egyptian-French J.V. (VINCI-Construction, Arab Contractors and ORASCOM Construction), Cairo, Egypt

Abstract. The roadway tunnels are considered a good solution for the success of any modern roadway network. Tunnels help greatly to overcome possible traffic congestion and considerably reduce journey time. The continuous growth of Traffic volumes leads to increased congestion and possibly decreased safety. This leads to requirements for better traffic flow including the provision of extra lanes and consequently extra tunnel space. The extra tunnel space can be achieved either by the widening of the existing tunnel or by adding a new one. The choice of the suitable method is dependent on many factors like tunnels alignment, site conditions, construction method, tunnel operation, risk assessment...etc.

The current research investigates the second alternative through a specific case study as an example. The method comprises adding two new tunnels to an existing twin roadway tunnels. The problem was investigated considering the new tunnels to be possibly added vertically or horizontally. The influence of the new tunnels construction on the existing tunnels is investigated considering both the variation of relative position and spacing distance in a parametric study context. Several numerical two and three-dimensional models are employed to check the construction sequence, the tunnelling safety and to evaluate the induced stresses in surrounding ground, straining actions in tunnels' liner and deformations of both ground and liner. The results help to find out the minimum practical and safe spacing distance between the driven new tunnels and the existing ones without the need to the relatively costly soil strengthening techniques.

Keywords: Roadway tunnels · Numerical modelling · Tunnels development

1 Introduction

Sometimes, adding new tunnels to existing roadway tunnels becomes the only feasible development method for existing tunnels when their traffic capacities are insufficient. New tunnels relative position, the spacing distance between new and existing tunnels and surrounding soil properties determine the impact of the new tunnels construction

S. Agaiby and P. Grasso (eds.), *Engineering Challenges for Sustainable Underground Use*, Sustainable Civil Infrastructures, DOI 10.1007/978-3-319-61636-0_6

on the existing ones. One of the main constraints during the new tunnels construction is to ensure the capability of the existing tunnels to operate safely.

The impact of addition new tunnels to existing tunnels has been investigated previously many times via site observation, physical models and numerical models [1–5]. In these studies, the attention was paid mainly to the induced deformation. Most of the applied numerical analyses were 2-D models with few attempts using 3-D models [6, 7]. A series of systematic 3-D analyses have been conducted to investigate the interaction between hypothetical large parallel twin tunnels constructed in stiff clay using the New Austrian Tunneling Method [8]. The analysis proved that for spacing distance larger than 3D (where D is the average tunnel diameter); there is no significant impact from tunnels on each other [9, 10].

The current research introduces a parametric study to study this influence in two different cases; vertically and horizontally added tunnels. Two and three-dimensional numerical models are implemented for this parametric study. The study considers the induced deformations in ground and tunnels and induced stress presented by plasticized zones in surrounding ground. The research aims mainly to find the minimum reliable spacing distance between driven new tunnels and existing ones without the consideration of soil strengthening. The results can be used to help the decision-making process as guidelines at the preliminary and layout design stage.

2 The Case Study

It is expected that adding new tunnels close to existing tunnels shall have an influence on ground deformations and stabilization of the existing tunnels. Even for Tunnel Boring Machine (TBM) technique, the tunneling process affects the existing neighbor tunnel lining during soil excavation and lining installation stages [11–13].

The current study is performed assuming the existence of twin two-lane roadway tunnels run parallel to each other in a shallow depth which is similar to El-Azhar roadway twin tunnels [14]. The existing roadway tunnels are located at distance 15.0 m from surface. These tunnels need development due to traffic capacity insufficiency. Thus, new two roadway tunnels shall be driven parallel to the existing ones using TBM method. All tunnels, existing and new, are assumed to have the same diameter (D) which is 9.4 m.

In this research, a parametric study considers the clear spacing distance (Ds) between new and existing tunnels is conducted. The investigated distance ranges from 0.1D to 1D. The parametric study considers, also, two different cases for tunnels relative positions, as shown in Fig. 1. The first case is to add the new tunnels (NT1 and NT2) vertically underneath existing tunnels (ET1 and ET2) to avoid the possible conflict with the existing utilities if the new tunnels will be considered to be constructed above the existing ones. The second case is to add them horizontally at the same elevation of existing tunnels. The parametric study is performed for cohesion-less sandy soil which commonly exists in Egyptian ground.

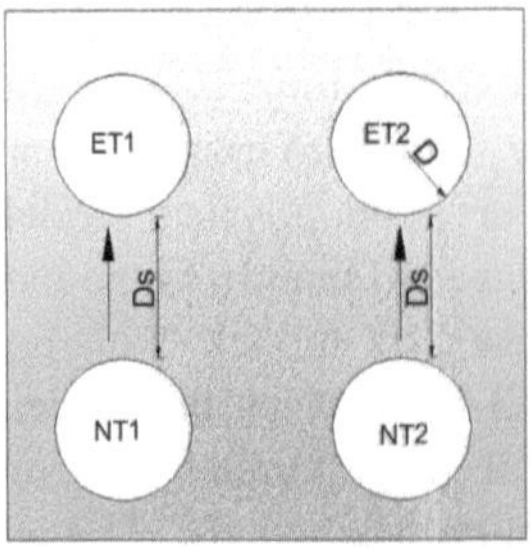

First case: Twin tunnels added vertically underneath existing tunnels

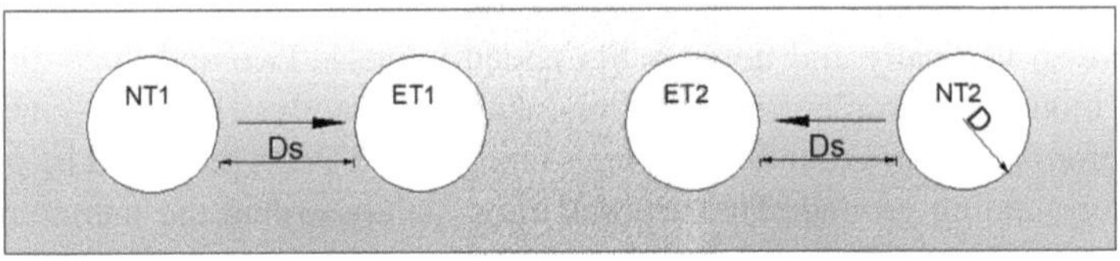

Second case: Twin tunnels added horizontally parallel to existing tunnels

Fig. 1. Tunnels relative positions

3 Numerical Modelling

Numerical simulation of TBM tunneling should consider many factors such as machine advance, shield stiffness, grouting pressure, slurry face pressure and lining installation. These factors should be considered in 3-D numerical modeling to get realistic simulation. In the 2-D analysis, some approximation can be performed to get reasonable simulation compared to 3-D analysis. In this research, 2-D plane strain numerical models, with elastoplastic soil model based on Mohr-Coulomb failure criterion, were developed for all study cases. 3-D modelling is employed for some critical cases. Figure 2 shows the mesh layout applied in 2-D models where the soil is represented by 3-node triangular elements and the tunnel lining is represented with 2-node elastic beam elements [15]. In the 3-D numerical model, 4-nodes tetrahedron solid elements are used to represent soil and tunnel lining and 4-nodes shell elements are used for shield modeling. The Finite Element Analysis package, MIDAS-GTS, is used for both 2-D and 3-D numerical modelling [16]. The main loading and construction stages considered in these models are:

1. Initial state: This considers the effect of soil pressure, surface load and ground water table.
2. Soil excavation for the first existing tunnel (ET1): A load distribution factor is used for excavated soil elements elimination.
3. Lining installation for ET1: The lining elements are activated after excavation of entire tunnel section.
4. Steps 2 and 3 are repeated for the second existing tunnel (ET2).
5. The long term effect.

6. Soil excavation for the first new added tunnel (NT1): Face pressure of the machine is applied and load distribution factor is used for excavated soil elements elimination.
7. Lining for NT1 activation: Lining and grouting elements are activated after entire tunnel soil excavation.
8. Steps 6 and 7 are repeated for the second new added tunnel (NT2).

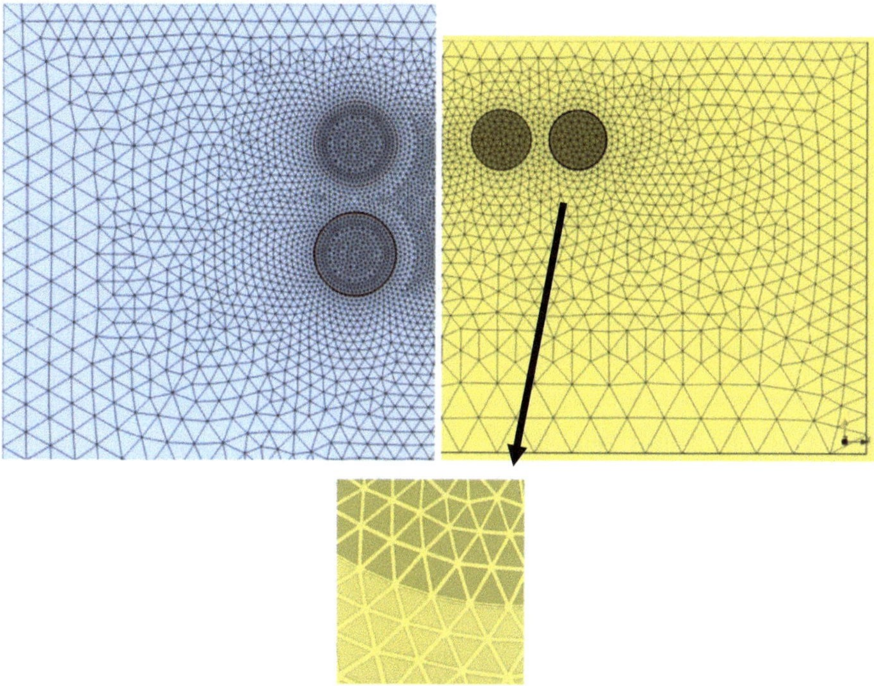

Fig. 2. Mesh layout for 2D numerical modeling of the two types of the parametric study

3.1 Material Properties

Tables 1 and 2 show the used characteristic material properties for existing tunnels lining, new tunnels lining, grout and ground soil respectively.

Table 1. Material properties of existing tunnels lining

Concrete properties	Thick. (m)	Density (γ) kN/m^3	Young's modulus (E) MPa	Poisson's ratio (ν)
Lining	0.4	25	28.5×10^3	0.2
Grout	0.125	20	150	0.25

Table 2. Soil parameters

	Model/parameter	Unit
Soft ground type	Sand	
Material model	Mohr-Coulomb	–
Type of material behavior	Drained	–
Soil unit weight (γ)	20	kN/m^3
Horizontal permeability (Kx)	1.0	m/day
Vertical permeability (Ky)	1.0	m/day
Young's modulus (E)	40000	kN/m^2
Poisson's ratio (ν)	0.37	–
Cohesion (C)	0.0	kN/m^2
Friction angle (Φ)	39	–

3.2 Analysis and Results

In this section, numerical models for each value in the investigated spacing distance range are established for the described relative position cases. Results, from these models, are plotted to show the influence of new tunnels construction on existing tunnels and ground surface deformations, and to show the plasticized zones distribution in soil and straining actions in tunnels lining.

3.3 Vertically Added Tunnels

3.3.1 Deformations of Existing Tunnels and Ground Surface

Figure 3 shows the impact of spacing distance variation on the ground surface settlement above tunnels centerline and the mid-point between tunnels. For all studied points,

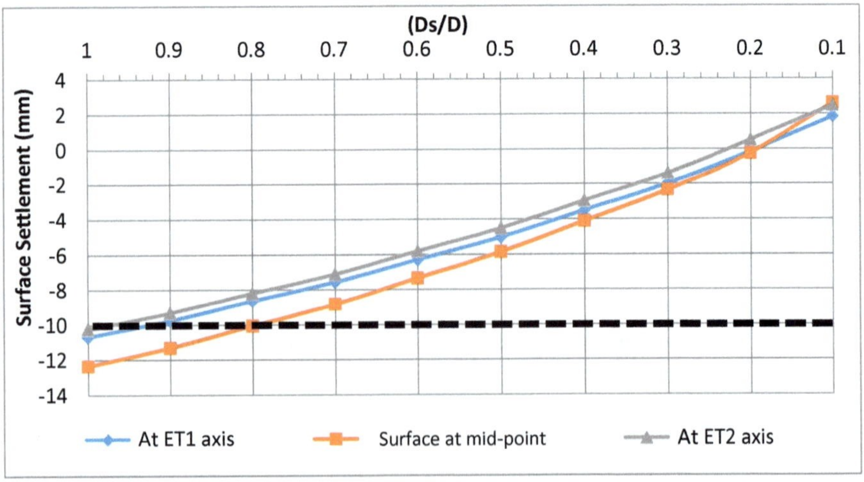

Fig. 3. Surface deformation due to (Ds) variation

the settlement value decreases as spacing distance decreases. The settlement values range from 10 to 12 mm at spacing distance 1D and reach 2 mm heave at spacing distance 0.1D. The existing tunnels lining deformations at crown and inverts are shown in Fig. 4. The deformation values range from 11 to 14.5 mm downward at a 1D spacing distance and decrease to reach values range from 1 mm downward to 2.5 mm upward at 0.1D spacing distance. Assuming that the preconditioned or allowed settlement value is 10 mm for the existing tunnels, and reviewing Figs. 3 and 4, the recommended spacing between the existing and the new tunnels is 0.5D.

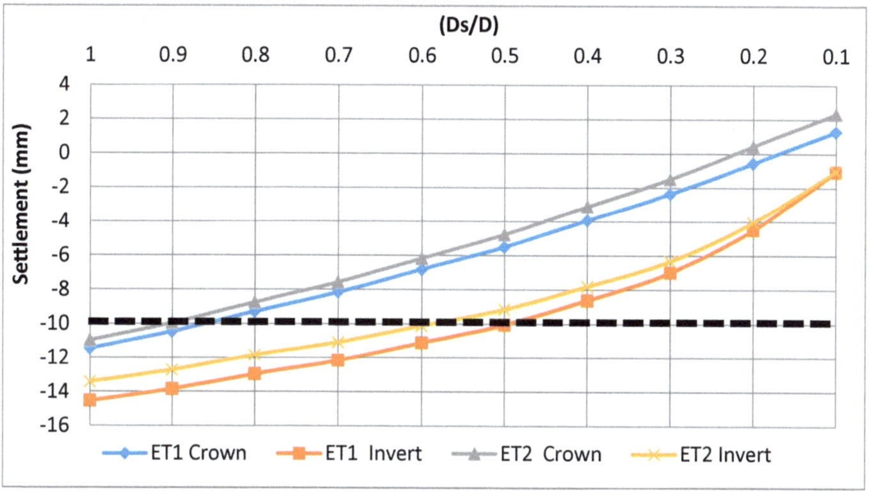

Fig. 4. Existing tunnel deformation due to (Ds) variation

3.3.2 Plasticized Zones Distribution in Soil

Plasticized zones distribution is one of the indications of soil failure probability due to excessive stresses. In the studied cases, this situation was found around existing tunnels due to nearby tunneling process. The behavior of confined soil between both of tunnels, existing and new ones, is either to be elasto-plastic or reach to fully plasticity. In case of plasticity distribution occupied whole confined soil height between existing tunnel and new one, the failure of existing tunnels is expected. In the developed 2-D numerical models, the plastic zone height is introduced as a percentage of the total spacing distance between existing and new tunnels which is shown in Fig. 5. The chart, shown in Fig. 6, shows that plasticized zone percentage increases with the decrement of the distance (Ds). For (Ds) equal to tunnel diameter D, the plasticized zone percentage distance is about 11% of the total spacing distance and this percentage is increased to be 95% at (Ds) equal to 0.1D which indicates soil failure.

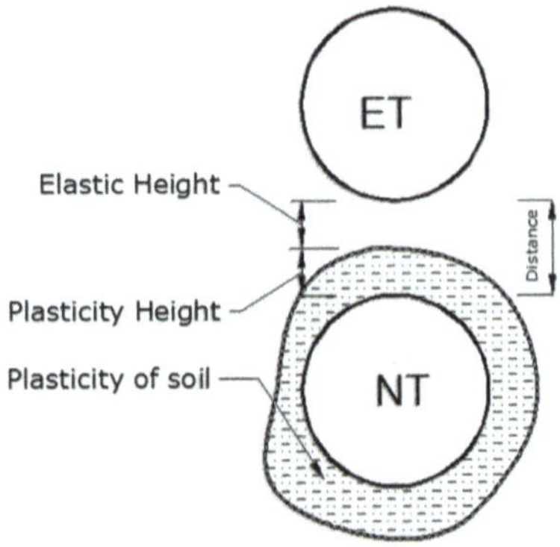

Fig. 5. Plastic and Elastic height between tunnels

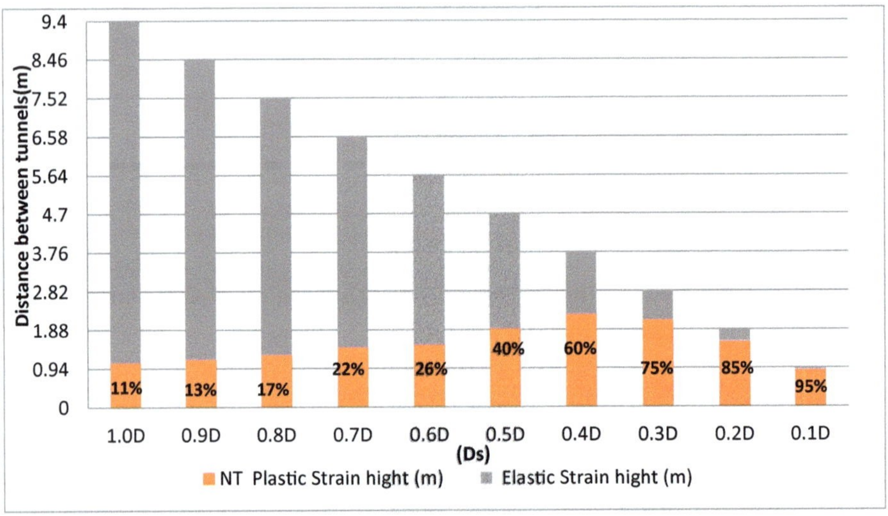

Fig. 6. Percentage of plasticized zone height to total spacing distance (Ds)

These results are compared and confirmed with two 3-D models for the spacing distance equal 0.1D and 0.5D. Figures 7 and 8 show the contour lines for plasticized zones distribution along the spacing distance in these models.

3.3.3 Straining Actions in Existing Tunnels Lining

For different spacing distance values, the maximum normal force value in existing tunnel (ET1) lining is reduced during first new tunnel (NT1) execution and increased during second new tunnel (NT2) execution. This trend is reversed for the maximum normal force value in the existing tunnel (ET2). The value is increased during first new tunnel (NT1) execution and decreased during second new tunnel (NT2) execution. These values are presented in bar charts shown in Figs. 9 and 10.

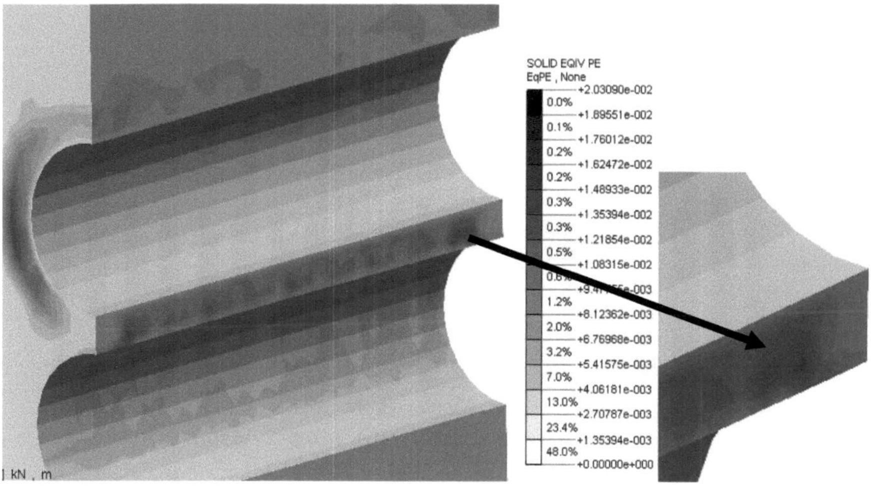

Fig. 7. Plasticized zone distribution at (Ds) equal to 0.1D

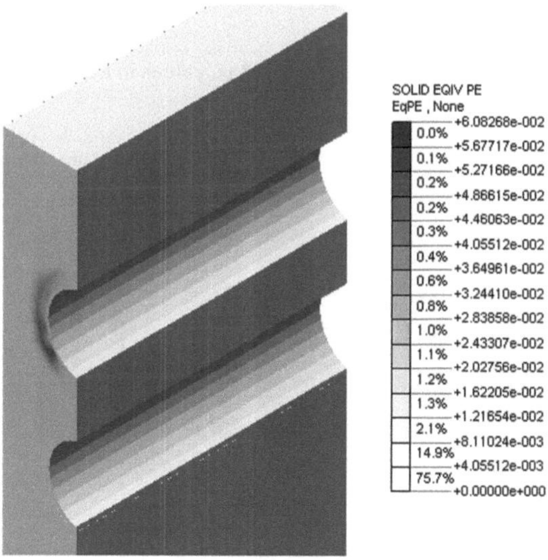

Fig. 8. Plasticized zone distribution at (Ds) equal to 0.5D

Induced bending moments' maximum value follow the same trend for induced normal forces in existing tunnels' lining, with an exception for small spacing distance cases. For the existing tunnel (ET1) with spacing distance 0.1D, the maximum bending moment value is increased during first new tunnel excavation and is reduced during lining installation. For the existing tunnel (ET2), the values variation, during new

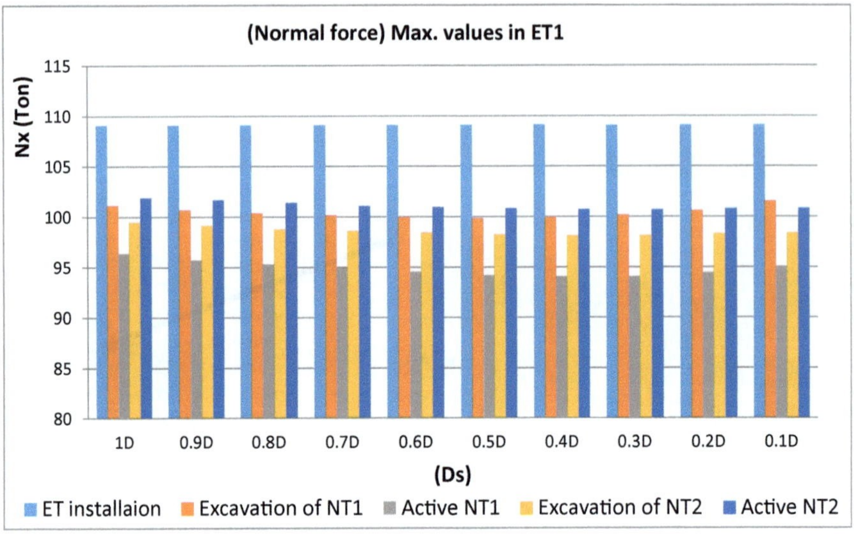

Fig. 9. Maximum normal force induced in existing tunnel ET1

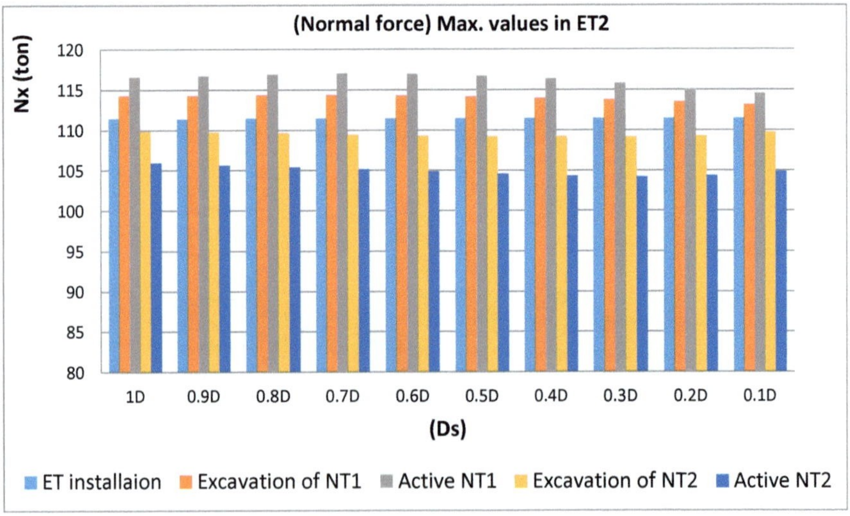

Fig. 10. Maximum normal force induced in existing tunnel ET2

tunnels execution, is reduced with spacing distance reduction. At spacing distances 0.2D and 0.1D, the value is almost settled during first new tunnel execution. These results are shown in Figs. 11 and 12.

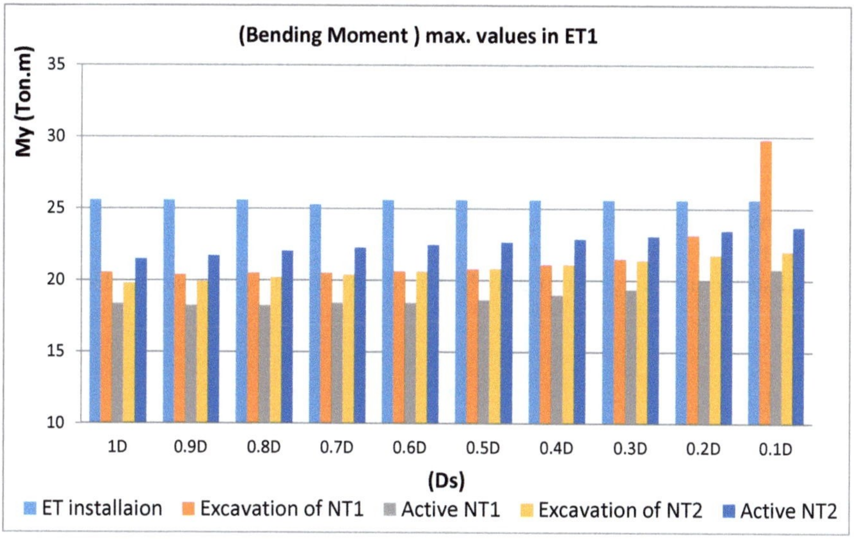

Fig. 11. Maximum bending moment induced in existing tunnel ET1

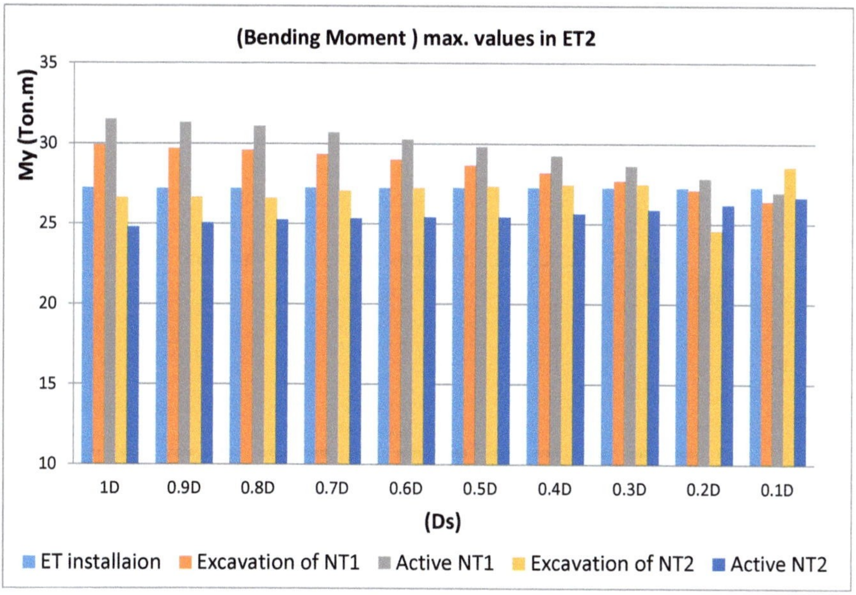

Fig. 12. Maximum bending moment induced in existing tunnel ET2

3.4 Horizontally Added Tunnels

3.4.1 Deformations of Existing Tunnels and Ground Surface

Figure 13 shows the surface deformation during new tunnels installation with different spacing distance. The deformation is plotted at five different points; above existing tunnels, at mid-point between existing tunnels and above new tunnels. The general behavior of the ground surface is heaving with peak values between mid-point and existing tunnel (ET1). The existing tunnels lining deformations at crown and inverts are shown in Fig. 14. The deformation values range from 8.5 to 9 mm upward at 1D spacing distance and range from 9 to 12.5 mm upward at 0.1D spacing distance.

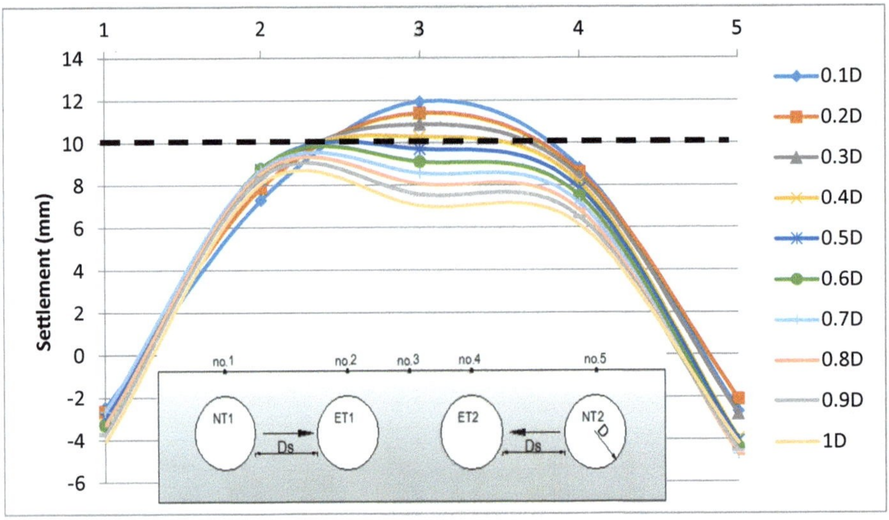

Fig. 13. Surface deformations due to (Ds) variation

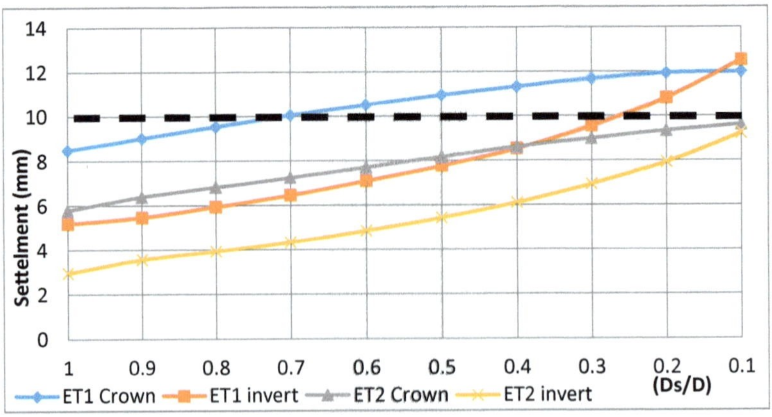

Fig. 14. Existing tunnels deformations due to (Ds) variation

3.4.2 Plasticized Zones Distribution in Soil

The plasticized zones distribution shows critical stability at the spacing distance equal 1D, as shown in Fig. 16. It also indicates probable soil failure for smaller spacing distances. An example is given in Fig. 15.

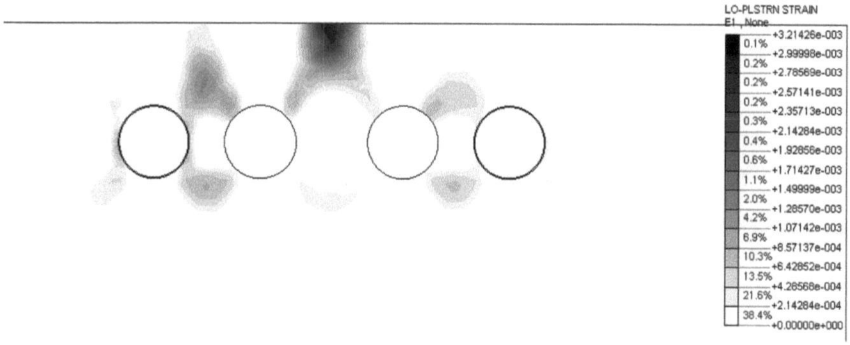

Fig. 15. Plasticized zones distribution at (Ds) = 0.5D

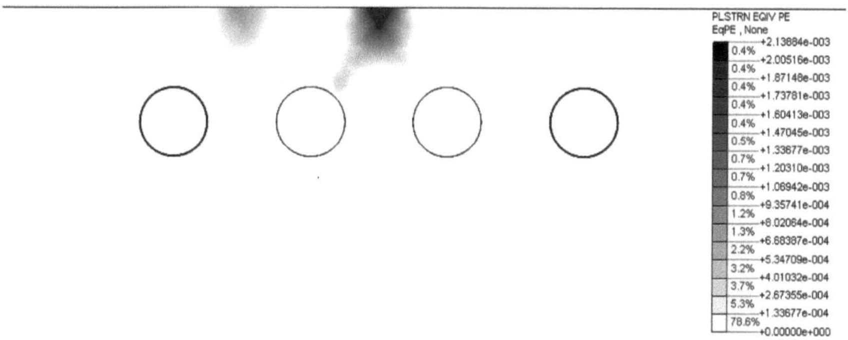

Fig. 16. Plasticized zones distribution at (Ds) = D

3.4.3 Straining Actions in Existing Tunnels Lining

For larger spacing distances (0.7D–1.0D), the induced normal forces maximum value, in ET1, is constant during new tunnels execution. For smaller spacing distances, the value is increased during execution. A significant increase occurs at a spacing distance of 0.1D, as shown in Fig. 17. This is the same behaviour for normal forces maximum values in ET2 due to new tunnel NT2 execution, as shown in Fig. 18.

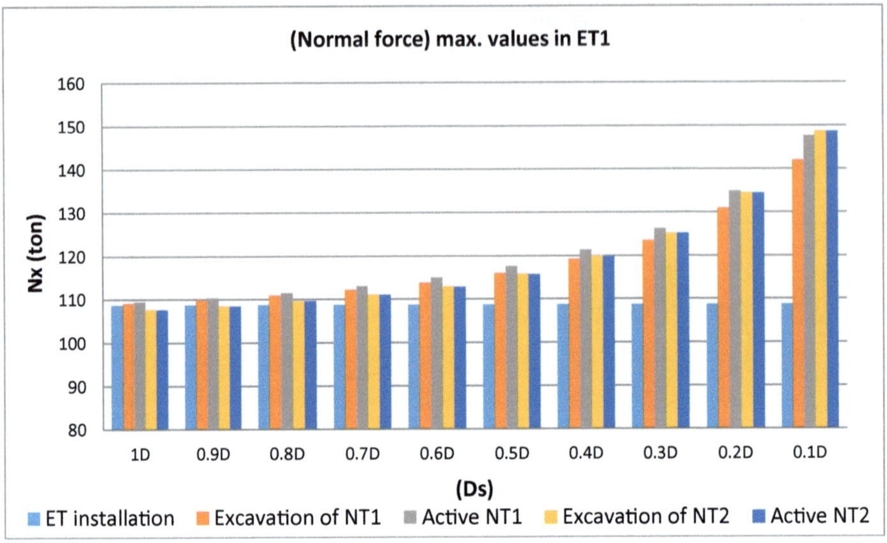

Fig. 17. Maximum normal force induced in existing tunnel ET1

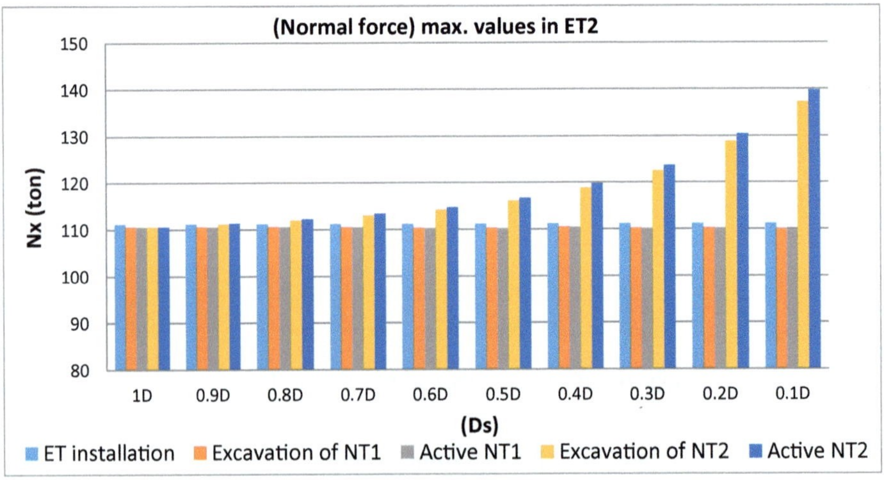

Fig. 18. Maximum normal force induced in existing tunnel ET2

In general, the induced bending moments' maximum value in existing tunnels lining is constant during new tunnels execution except for small spacing distances (0.1D–0.3D), Figs. 19 and 20.

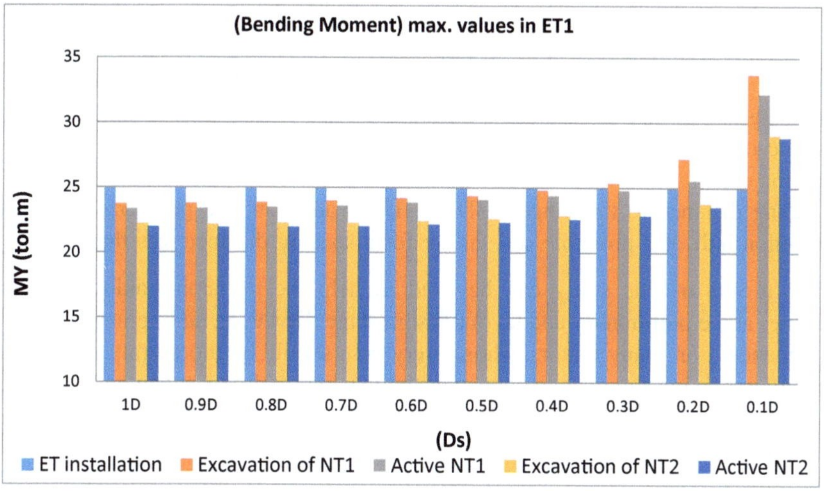

Fig. 19. Maximum bending moment induced in existing tunnel ET1

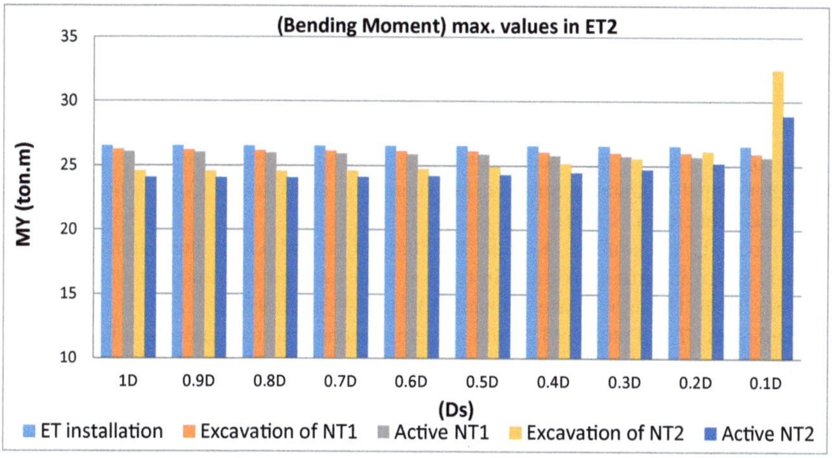

Fig. 20. Maximum bending moment induced in existing tunnel ET2

4 Conclusions

For the investigated cases study, defined with soil type, tunnelling depth and tunnels diameter, the following conclusions may be drawn:

- The process of adding new tunnels generally does not affect the straining actions induced in existing tunnels lining beyond the safety limits. For all studied cases, lining internal stresses due to induced straining actions did not exceed the material allowable stress. Thus, existing lining internal stresses may not be considered as a constraint in spacing distance determination.

- In the case of vertically added tunnels, due to preconditioned deformation constraint, the recommended spacing distance is 0.5D. Attention should be paid to soil plastic zones in addition to settlement value to choose the recommended spacing. Spacing can be further reduced with the use of soil strengthen technique.
- In the other case of horizontally added tunnels and due to preconditioned deformation constraint, it is recognized that existing tunnels invert deformation limit the spacing distance to the range 0.3D to 1D. In addition, the plasticized zones distribution indicates probable soil failure for a spacing distance less than 1D.

The results of this study can be used to help the decision-making process as guidelines at the preliminary and layout design stage. However, adding new tunnels is a case dependent on many factors. Thus, every case should be thoroughly analysed and checked after fixing the distance between the existing and new tunnels.

References

1. Kim, S.H., Burd, H.J., Milligan, G.W.E.: Model testing of closely spaced tunnels in clay. Geotechnique **48**(3), 375–388 (1998)
2. Liu, H.Y., Small, J.C., Carter, J.P., Williams, D.J.: Effects of tunnelling on existing support systems of perpendicularly crossing tunnels. Comput. Geotech. **36**(5), 880–894 (2009)
3. Karakus, M., Ozsan, A., Basarir, H.: Finite element analysis for the twin metro tunnel constructed in Ankara Clay-Turkey. Bull. Eng. Geol. Environ. **66**, 71–79 (2007). doi:10.1007/s10064-006-0056-z
4. Perri, G.: Analysis of the effects of the new twin-tunnels excavation very close to a big diameter tunnel of Caracas Subway. In: Tunnelling and Ground Conditions, pp 523–530. Balkema, Rotterdam (1994)
5. Yamaguchi, I., Yamazaki, I., Kiritani, K.: Study of ground-tunnel interactions of four shield tunnels driven in close proximity, in relation to design and constructions of parallel shield tunnels. Tunn. Undergr. Space Technol. **13**(3), 289–304 (1998)
6. Addenbrooke, T.I., Potts, D.M.: Twin tunnel interaction: surface and subsurface effects. Int. J. Geomech. **1**(2), 249–271 (2001)
7. Koungelis, D.K., Augarde, C.E.: Interaction between multiple tunnels in soft ground. School of Engineering, University of Durham, UK (2004)
8. Lee, K.M., Ng, C.W.W., Tang, D.K.W.: Three-dimensional numerical investigations of New Austrian Tunnelling Method (NATM) twin tunnel interactions. Can. Geotech. J. **41**, 523–539 (2004)
9. Hage, C.F., Shahrour, I.: Numerical analysis of the interaction between twin-tunnels: influence of the relative position and construction procedure. Tunn. Undergr. Space Technol. **23**(2), 210–214 (2008)
10. Chakeri, H., Hasanpour, R.: Analysis of interaction between tunnels in soft ground by 3D numerical modeling. Bull. Eng. Geol. Environ. **70**, 439–448 (2010, 2011). doi:10.1007/s10064-010-0333-8
11. Abdou, M., El-Wafa, W.A.: Behavior of existing tunnel due to the construction of a new tunnel passed parallel under it. J. Eng. Sci. Assiut Univ. **35**(6), 1381–1400 (2007)
12. Ezzeldine, O.Y., Darrag, A.A.: Instrumentation at the CWO crossing El-Azhar road tunnels and its use in the design of future projects. In: International Symposium on Utilization of Underground Space in Urban Areas, Sharm El-Sheikh, Egypt, 6–7 November 2006

13. Karakus, M., Fowell, R.J.: 2-D and 3-D finite element analyses for the settlement due to soft ground tunnelling. In: ITA-AITES 2006 World Tunnel Congress and 32nd ITA General Assembly: Proceedings Safety in the Underground Space, vol. 21, p. 392 (2006)
14. Abu-Krisha, A.: New 2D simulation of TBM tunneling and application on EL-Azhar road tunnels. In: 28th, ITA-AITES World Tunnel Congress, Sydney, Australia (2005)
15. Zaki, M., Abu-Krisha, A.: Numerical simulation of TBM and NATM for Cairo Metro Line 3. In: Proceedings of the AITES-ITA World Tunnel Congress, Utilization of Underground Space in Urban Areas 2006, Sharm El-Sheikh, Egypt (2006)
16. MIDAS, MIDAS GTS NX V3.0.0(4): MIDAS GTS Manual. MIDAS Information Technology Co., Ltd., Gyeonggi 463-400, Korea

Estimation of Settlement and Vibration on the Surface Due to the Construction of East-West Metro Tunnels in Kolkata, India

Aniruddha Sengupta[1(✉)], Raj Banerjee[1,2], and Srijit Bandyopadhyay[1,2]

[1] Indian Institute of Technology, Kharagpur 721302, India
sengupta@civil.iitkgp.ernet.in
[2] BARC, Mumbai, India

Abstract. Kolkata, located on the bank of River Ganges in India, is considered to be one of the most densely populated cities in the world. It has an underground metro rail connection cutting across the city running predominantly in the north-south direction. Another underground metro rail connection is being built in the east-west direction connecting the heart of the city to the railway stations and the airport. As a part of the project, twin tunnels of diameter 6.1 m and 15 m apart are being constructed below the Ganges and through the most congested part of the city. The tunnels are located at a depth between 17 m and 24 m below the ground surface. The soil layers in Kolkata essentially consists of soft clays (clayey silt to silty clay) deposited by the river with layers of dense sand at much greater depth. Several old heritage structures have been identified along the route of the E-W metro. These structures are located between 19 m and 33 m from the centerline of the tunnels. Damage to these heritage buildings during the construction of the tunnels due to excessive settlement and/or vibration is a concern. Both static and dynamic finite element analyses have been performed to estimate the settlement and the vibration at those heritage structures due to the tunnel constructions. The results of the analyses are compared with the reported case histories of tunnel constructions worldwide on soft soils and with the empirical methods proposed by Mair et al. and FHWA to estimate the ground settlement due to the proposed tunnel construction.

1 Introduction

Kolkata, the capital of the Indian state of West Bengal, is located on the bank of River Ganges in India. This cosmopolitan city is considered to be one of the most densely populated cities in the world. It has an existing underground metro rail connection cutting across the city running predominantly in the north-south direction. Another underground metro rail connection (E-W metro) is being built in the east-west direction connecting the business district (Brabourne Road) of the city to the main railway station (Howrah Station) across the River Ganges and the international airport. Figure 1 shows the route of the E-W metro. As a part of the project, twin tunnels of 6.1 m in diameter and 15 m apart (centre to centre) are being constructed below the River Ganges and through the most congested part (Brabourne Road) of the city. Figure 2

S. Agaiby and P. Grasso (eds.), *Engineering Challenges for Sustainable Underground Use*, Sustainable Civil Infrastructures, DOI 10.1007/978-3-319-61636-0_7

shows a ground view of the Brabourne Road, the main business district of Kolkata. The tunnels are located at a depth between 17.6 m and 23.7 m below the ground surface (KMRC 2015). Several old heritage masonry structures have been identified along the route of the E-W metro. These heritage structures are located along the Brabourne Road/JN Road and shown by a circle in Fig. 1. These structures are located between 22 m and 33 m from the centerline of the tunnels. All most all of these old heritage masonry structures are on raft foundation with the bottom of the raft located about 2 m below the ground level. Damage to these heritage buildings during the construction of the tunnels due to excessive settlement and/or vibration is a major concern. The newly constructed high rise concrete reinforced buildings located along the route of the E-W metro are on pile foundations and considered to be less susceptible to damage due to the tunnel construction. A study is undertaken to estimate the ground settlement and the vibration at the ground level during the construction of the tunnels. This paper summarizes the outcome of this investigation.

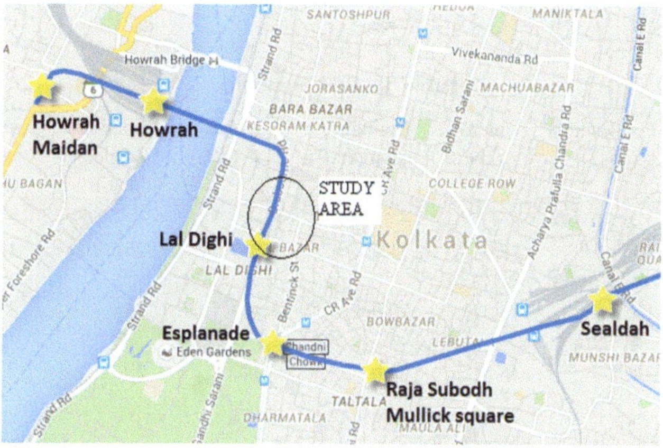

Fig. 1. E-W metro route & the study area.

Fig. 2. Ground view of braboune road (study area).

2 Subsoil Layers and Their Properties

The subsoil layers in the vicinity of the E-W metro and the heritage structures in Kolkata are obtained from the borehole logs and essentially consist of soft clays (clayey silt to silty clay) deposited by the river with layers of dense sand at much greater depth. As per the borehole logs, the top 2 m is fill material with a permeability (k) of 5×10^{-5} m/s. The ground water table is located at 1 m below the ground surface during rainy season. From 2 m to 15.4 m, the soil is silty clay or clayey silt (Unit 2). From 15.4 m to 19.5 m, the soil is medium silty clay (Unit 3a). Below 19.5 m, the soil is silty clay (Unit 3b) with some sand. The soil strength parameters for the soils are shown in Table 1. The value for Poisson's ratio of the soils (v_s) is assumed to be 0.3. The deformation modulus of a soil is determined from the well established correlation (USACE 1990) between the deformation modulus (Es), the undrained strength of the soil (C_u), overconsolidation ratio (OCR) and plasticity index (PI) of the soil, as shown below.

$$E_s = K_c * C_u \quad (1)$$

Table 1. Soil properties.

Soil type	Permeability k (m/s)	Unit Wt., γ (kN/m^3)	Undrained strength, Cu (kPa)	Cohesion, c′ (kPa)	Friction Angle, ϕ' (deg.)	Deformation modulus Es (kPa)
Fill	5×10^{-5}	–	–	–	–	–
Silty clay (Unit 2)	10^{-7}	18.5	31.0	1.0	25	18600
Medium silty clay (Unit 3a)	10^{-9}	18.5	70.0	0.0	31	31500
Silty clay (Unit 3b)	10^{-6}	18.5	55.0	0.0	29	27500

The values for the parameter K_c is obtained from Fig. 3 based on the PI and OCR for a soil. The soils are conservatively assumed to be normally consolidated (OCR = 1) in this study. The average plasticity index for the Units 2, 3b and 3b soils are 30, 45 and 35, respectively. Accordingly, the values of K_c for these soils are assumed to be 600, 450 and 500, respectively from the chart. The coefficient of earth pressure at rest is assumed to be 0.5 for all the soils. The average undrained strength of these soils are 31, 70 and 55 kPa, respectively based on the laboratory tests. The strengths (γ = unit weight of the soil, c′ = effective cohesion of the soil and ϕ' = effective friction angle of the soil) of the fill materials at the top is assumed to be same as that of the silty clay/clayey silt (Unit 2) layer except the value of the permeability.

3 Settlement Due to E-W Tunnels Construction

The alignment of the proposed twin tunnels with respect to the three identified heritage structures indicates that one of the heritage structures (Currency Building) is 33 m away from the rightmost tunnel while the other two (Bethel Synagogue and Maghen

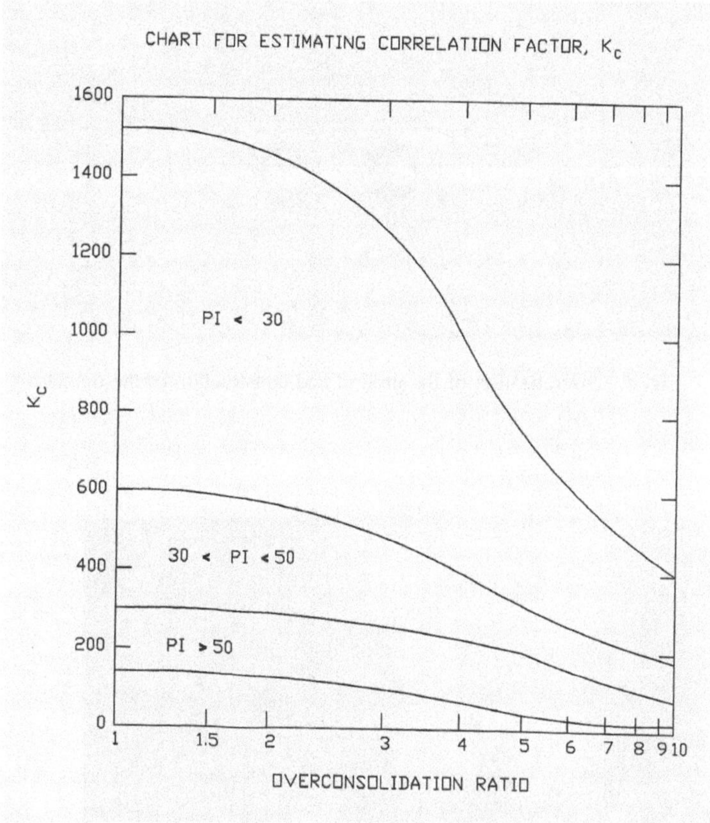

Fig. 3. Estimation of K_c based on OCR and PI of a soil (Ref. USACE 1990).

David Synagogue) are 22 m and 23 m away from the centerline of the second (rightmost) tunnel (KMRC 2015). The twin tunnels are 15 m apart (centerline to centerline). The tunnels are 6.1 m in diameter. The width of the concrete (M40) lining is about 350 mm. The deformation modulus and Poisson's ratio of the concrete are assumed to be $E_{conc} = 3.162E + 07$ kPa and $\upsilon_{conc} = 0.15$, respectively. The proposed crown (top) depth and invert (bottom) depth of the tunnels at the study location are 17.6 m and 23.7 m, respectively. The twin tunnels shall be constructed by earth pressure balance shield (EPBS) method. Based on KMRC (2015), it is assumed that the tunnels are not constructed at the same time. The second tunnel shall be constructed when the first tunnel is constructed and advanced by at least 100 m.

The numerical analyses are performed using PLAXIS 2D software (Plaxis bv, 2012). PLAXIS2D is based on two dimentional plane strain finite element method and it is suitable for analyzing any coupled soil-structure interaction problem where large deformation is envisaged. The two extreme side boundaries are located 50 m from the edge of the tunnels and they are considered to be on roller, that is, only vertical movement is allowed at these boundaries. The bottom boundary is located 20 m below

Fig. 4. Discretization of the ground and construction of the tunnels.

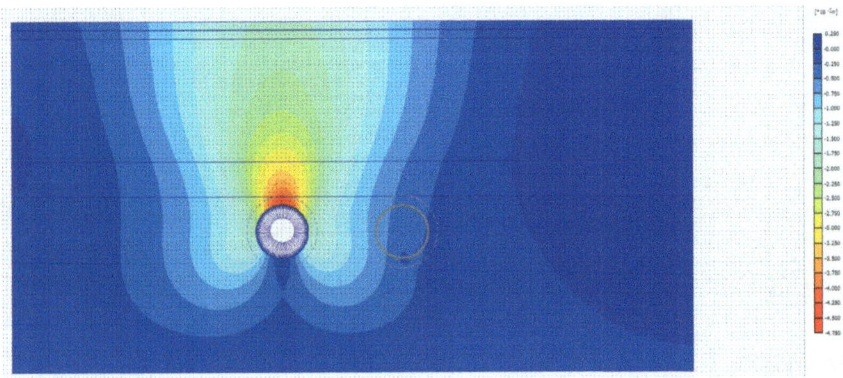

(a)

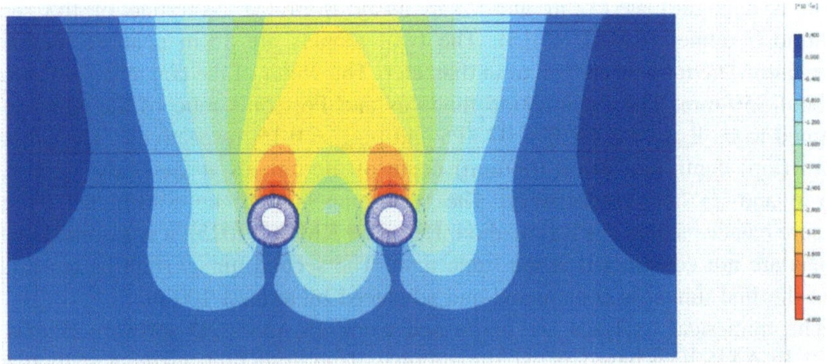

(b)

Fig. 5. Settlements due to the construction of (a) 1st tunnel and (b) 2nd Tunnel.

the bottom of the tunnels and only horizontal movement is allowed at this boundary. The soil is discretized by 6 noded triangular/8 noded rectangular isoparametric elements. The soils are assumed to be following the nonlinear elasto-plastic behavior satisfying the Mohr-Coulomb yield criteria. The tunnel linings are assumed to be elastic, impervious and seepage loss during tunnel construction is assumed to be minimal. The grouting and grout loss around the tunnels are not considered directly in the numerical analyses. Two cases with respect to the volume loss due to the construction of the tunnels (V_L) are considered. In the first case, V_L is assumed to be 2% as in FHWA (2009). This value corresponds to poor practice during tunnel boring with closed face machine in raveling ground. In the second case a V_L of 0.25% is assumed. This case represents good practice followed with tight control of face pressure within closed face machine in slowly raveling or squeezing ground. This value is found to yield observed settlements in tunnel construction with earth pressure balance (EPB) method in soft marine clay in Singapore (Gouw, 2005). In all the numerical analyses, the bottoms of the tunnels are assumed to be fixed, that is, they are not allowed to readjust their positions during construction. Note that the buildings or their foundations located in the area are not numerically modeled in these analyses. The effect of building weights and their stiffness are not considered in these analyses. However, the differential settlements at those places are obtained to perform more detail structural analyses of the buildings which are not part of this paper. At the initial stage of the numerical analyses, the subsurface without the tunnels is assumed to be in equilibrium condition. After this, the two tunnels are constructed one by one. After the construction of each tunnel, the stresses and deformations are obtained. Figure 4 shows the numerical discretization of the ground and the construction of the two tunnels. Figure 5 shows the settlements due to the construction of the first and second tunnels. Here the vibration due to the TBM operation is neglected. Figure 6 summarizes the ground surface settlement due to the construction of the first and second

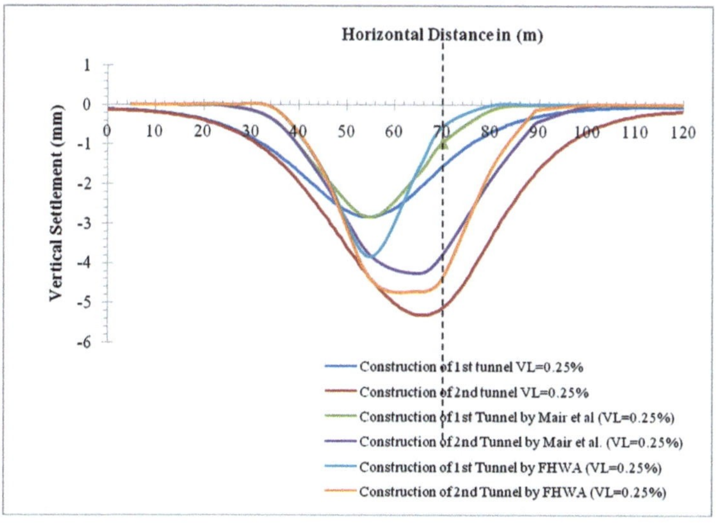

Fig. 6. Ground settlement after construction of both tunnels with $V_L = 0.25\%$.

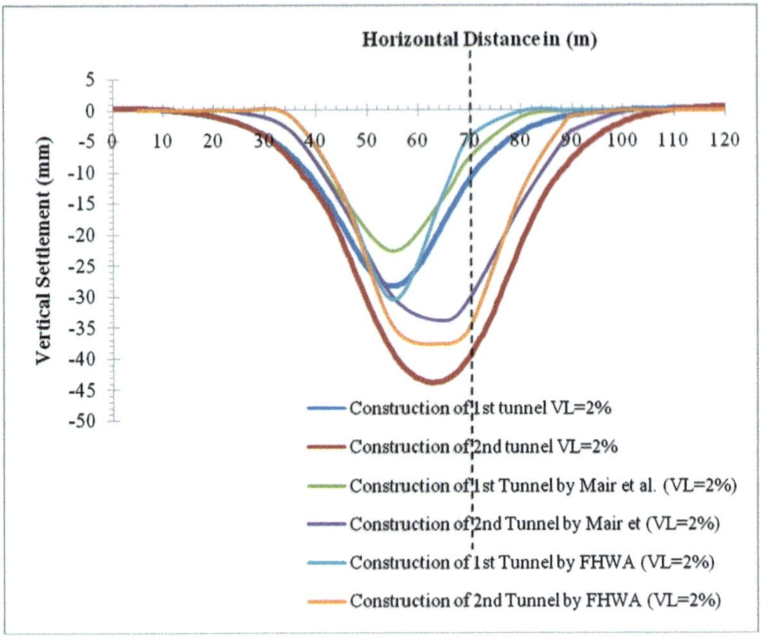

Fig. 7. Ground settlement after construction of both tunnels with V_L = 2%.

tunnel at the Brabourne Road area. In Fig. 6, the volume loss during tunnel construction is assumed to be 0.25% for all the cases. The finite element results are compared with the existing empirical methods given by Mair et al. (1996), and FHWA (2009). Figure 7 shows the estimated ground settlement due to the construction of first and second tunnels with the volume loss during tunnel construction is assumed to be 2% for all the cases. Here also, the finite element results are compared with the existing empirical methods given by Mair et al. (1996), and FHWA (2009). The estimated total settlement and angular distortion at the heritage structures at the end of the second tunnel construction are shown in Table 2. The estimated angular distortions of the buildings are well within

Table 2. Estimated total settlement and angular distortion due to tunnels construction.

Heritage structure	Distance from CL of 2nd tunnel (m)	Assumed volume loss (V_L)	Estimated maximum vertical settlement (mm)	Estimated maximum angular distortion
Currency Building	33 m	0.25	0.6	1/50000
		2.0	1.2	1/6087
Maghen David Synagogue	23 m	0.25	1.3	1/10000
		2.0	5.5	1/1650
Bethel Synagogue	22 m	0.25	1.4	1/10000
		2.0	6.0	1/1600

the permissible limit of 1/750. Note that it makes a significant difference in terms of settlement if you are on the left side of the 1st tunnel or if you are on the right side of the 2nd tunnel. Since it is not known at this time which tunnel will be first constructed, all the structures are conservatively assumed to be located on the right side of the 2nd tunnel.

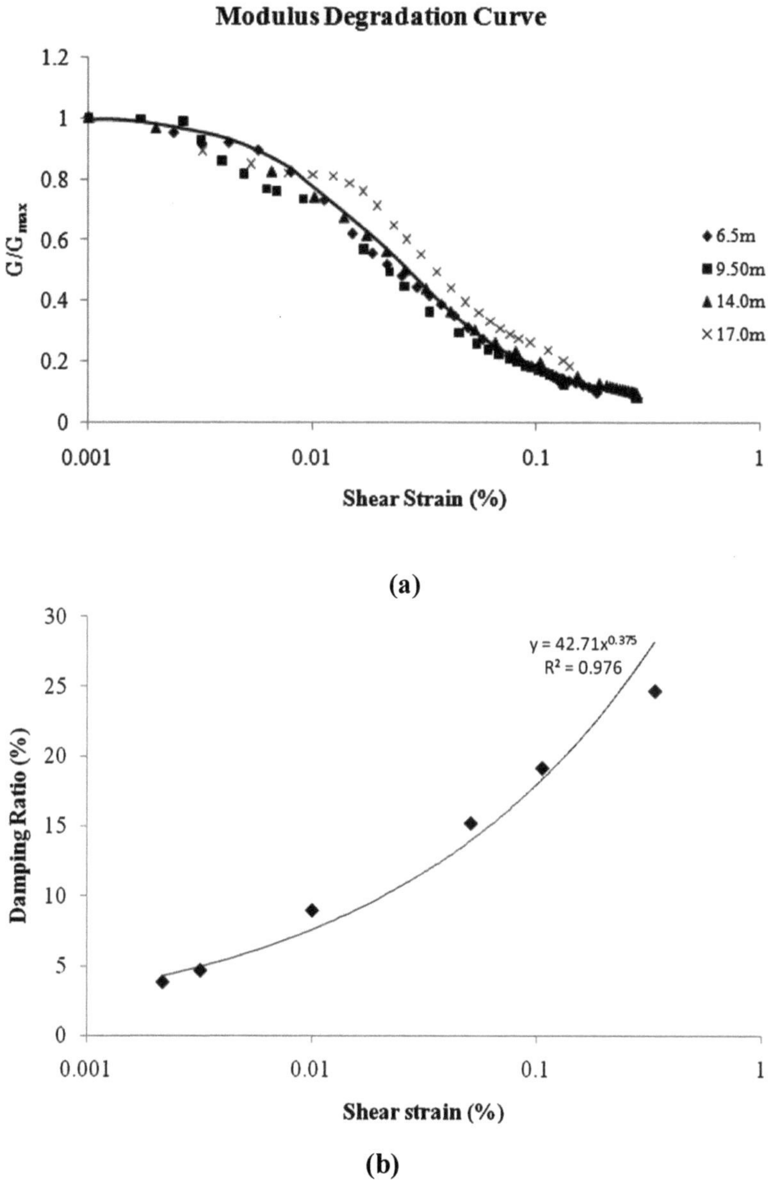

Fig. 8. Dynamic properties of kolkata soils.

4 Effect of Vibrations Due to TBM Operation

A dynamic analysis is followed after the construction of each tunnel to estimate the effect of vibration on the ground and foundation of the nearby structures. Here again, the structures on the ground are not modeled. Absorbing boundary conditions are assumed at all the boundaries to prevent reflection/refraction of waves from the boundaries. The dynamic properties of the soils are obtained from the cyclic triaxial tests performed on the Kolkata subsoils. The dynamic properties of the soils (for all the units) are given by one single shear modulus degradation (G/G_{max}) curve and one single damping ratio (β/β_{crit}) curve as shown in Fig. 8. In absence of any data on the vibrations generated by TBM in Kolkata subsoil and with reference to the motions given by Mooney et al. (2014), a white Gaussian vibrational motion is generated using random number. Figure 9 shows the generated vibrational motion which is utilized to model the ground vibration to be generated by the TBM in the Kolkata subsoils. This noise is applied radially to the tunnel linings during their construction and their effects on the ground surface are obtained. Note that vibrations are not applied simultaneously to both the tunnels as they are constructed one after another with a gap in between. Figure 10 shows the responses (motions) coming to one of the heritage buildings called Currency Building (located 33 m from tunnel) due to the construction of the E-W tunnels. The maximum amplitude of vibration is 0.002 m/s^2 at the Currency Building and 0.003 m/s^2 at the other two heritage buildings, Bethel Synagogue and Meghen David Synagogue. Figure 11 shows the vertical settlement profiles due to the TBM vibrations during the construction of the first and the second tunnels at the Brabourne Road area. Figure 12 shows the settlement profiles after the tunnel constructions. The vertical settlement at the Currency Building due to the vibration caused by TBM during the construction of the twin tunnels is estimated to be 0.009 mm. The vertical settlement at the Bethel and Meghen David Synagogues due to the vibration caused by TBM during the construction of the twin tunnels is estimated to be 0.01 mm from the above analyses. Both of these numbers are very small for further consideration.

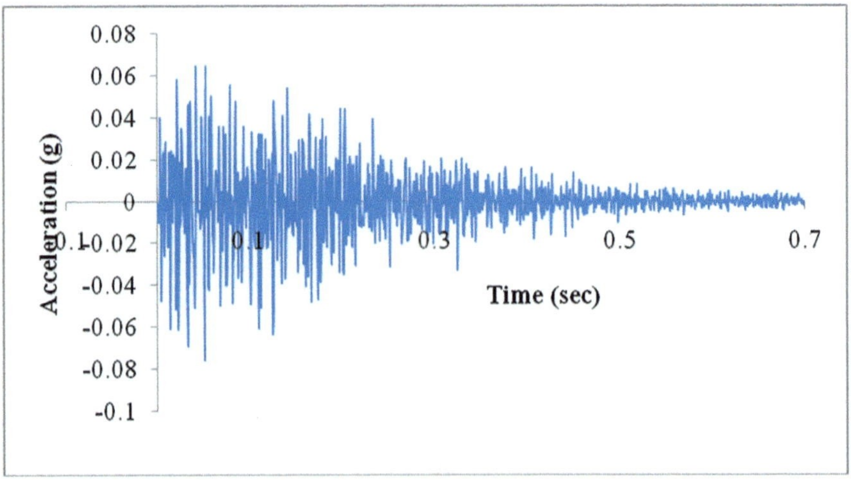

Fig. 9. Generated white gaussian noise to model vibration due to TBM operation in a tunnel.

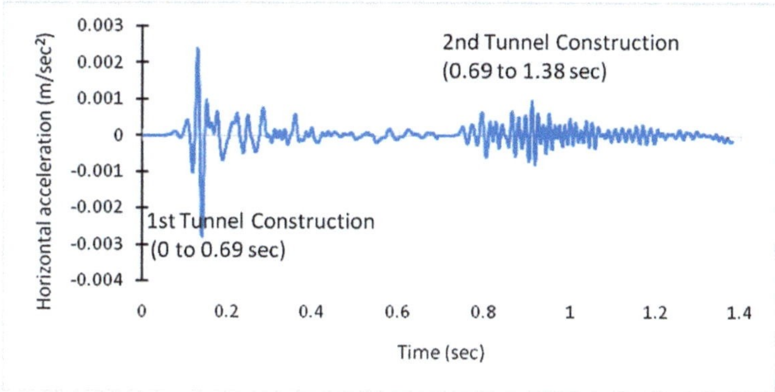

Fig. 10. Horizontal response at currency building due to tunnel construction.

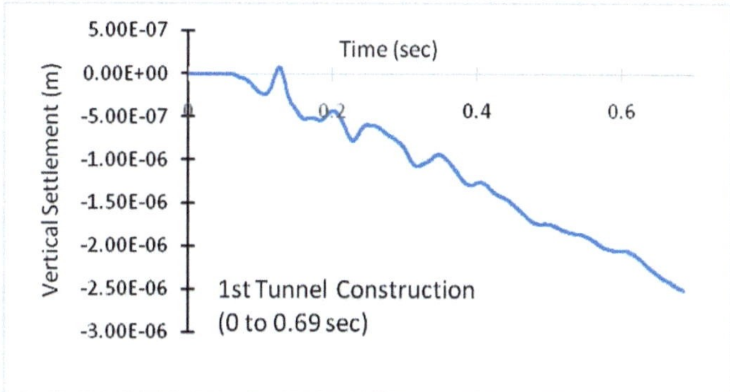

(a)

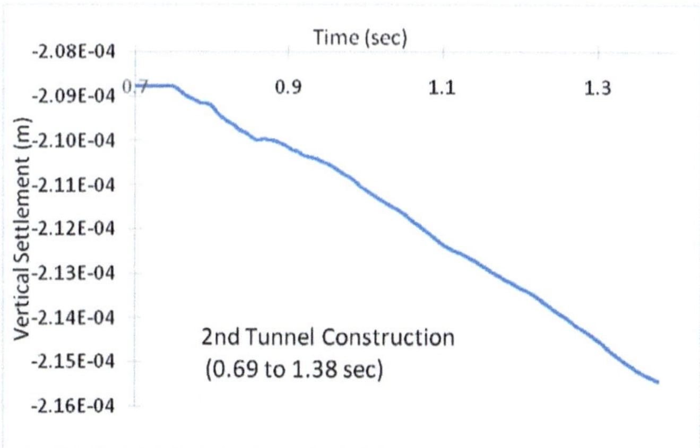

(b)

Fig. 11. Vertical settlement at currency building due to TBM noise during construction of the tunnels.

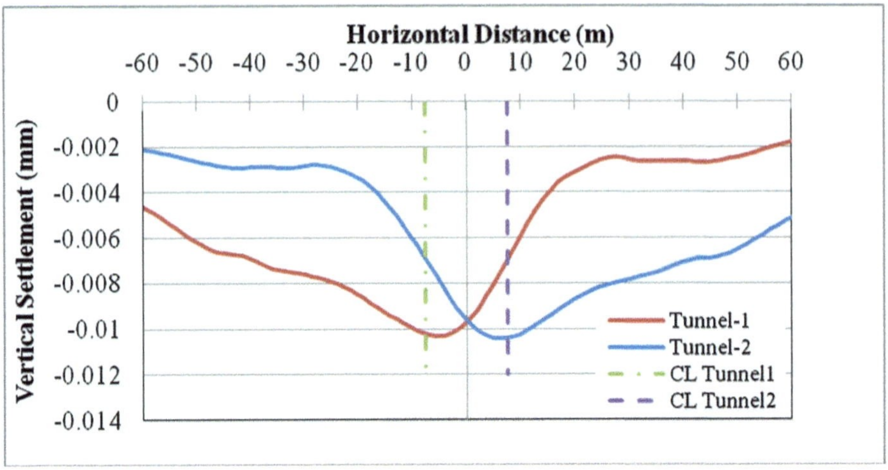

Fig. 12. Ground surface settlement profile after construction of tunnels

5 Conclusions

The vertical settlements due to the vibrations during the TBM operations are found to be negligible from the analyses. However, actual vibrations generated during TBM operation at the Kolkata subsoil needs to be determined to accurately determine their effect on the structures above the ground. The vertical settlement due to the tunnel constructions at the Currency building (located 33 m from the tunnel) is estimated to be between 0.6 to 1.2 mm. The angular distortions are found to be negligible. The vertical settlements at the Bethel and Meghen David Synagogues, located between 22 m and 23 m, are estimated to be between 1.3 and 6 mm. The maximum angular distortion is estimated to be 1 in 1600. This is well within the permissible limit of 1 in 750 for the residential buildings.

Acknowledgments. The funding and cooperation received from Kolkata Metro Rail Corporation during this study is hereby acknowledged.

References

FHWA: Technical Manual for Design and Construction of Road Tunnels – Civil Elements, FHWA-NH1-10-034, U.S. Department of Transportation, Federal Highway Administration, Washington DC, pp. 7–20, December 2009

Gouw, T.-L.: Tunneling Induced Ground Movement and Soil Structure Interactions. Seminar on Tunnel Technology in Civil Engineering, Peninsula Hotel, Jakarta, Indonesia, pp. 1–13, 22 March 2005

KMRC: Application for Permission for Construction Work of Mahakaran Underground Station and Tunnel near the Protected Monument of Currency Building along with Assessment Study Report. Kolkata Metro Rail Corporation, June 2015

Mair, R.J., Taylor, R.N., Burland, J.B.: Prediction of ground movements and assessment of risk of building damage due to bored tunneling. In: Mair, R.J., Taylor, R.N. (eds.) Geotechnical Aspects of Underground Construction in Soft Ground, pp. 713–718. Balkema, Rotterdam (1996). ISBN 90 5410 856 8

Mooney, M., Walter, B., Steele, J., Cano, D.: Influence of geological conditions on measured TBM vibration frequency. In: Davidson, G., et al. (eds.) North American Tunneling Proceedings, pp. 397–406. SME, Colorado (2014)

Plaxis bv: PLAXIS2D, Delft, Netherlands (2012)

USACE: Engineering and Design: Settlement Analysis. CECW-EG EM1110-1-1904. Department of the Army, U.S. Army Corps. of Engineers, Washington DC, 20314-1000, 30 September 1990

Effect of Using TBM Tunneling in Granular Soils on Performance of an Existing Raft Foundation

Ahmed M. El Mouchi[1(✉)], Asmaa M. Hassan[2], and Mohamed I. Amer[2]

[1] Faculty of Engineering, Cairo University, Giza, Egypt
ahmed.Elmouchi@dar.com
[2] Public Works Department, Faculty of Engineering, Cairo University, Giza, Egypt
{asmaa.moddather,m.amer}@cu.edu.eg

Abstract. The design of a tunnel in an urban area requires a proper estimate of the effect of tunnel construction on adjacent structures. Therefore, a 3D numerical modeling technique is necessary to understand the performance of shallow foundations during and after the construction of a nearby tunnel. A verification model has been used to validate the tunnel modeling and its output using field data measurements. The 3D finite element model has been used to evaluate the performance of a raft foundation in the presence of a tunnel. In addition, the effect of related parameters such as; soil relative density, tunnel diameter, tunnel cover, and horizontal clearance between the raft foundation and the tunnel have been studied. The settlement of the raft and the induced bending moments during the process of tunnel advancement are presented.

1 Introduction

Underground transportation systems have become a necessity in urban areas where traffic is congested and limited spaces are available. These systems require to construct tunnels underneath or adjacent to existing structures. Due to the interaction between tunnels and existing structures (Mroueh and Shahrour 2003; Maleki et al. 2011), the design and construction of tunnels especially in soft ground and at shallow depths is a major challenge. The prediction of additional induced settlements and straining actions due to tunneling has become a key issue in the planning process.

In previous research, the effect of tunneling on existing structures was predicted using a combination of in-situ observations and numerical modeling. Peck (1969) used the measured data from greenfield sites to obtain simple empirical equations. Similar relationships were suggested by other researchers (O'Reilly and New 1982; Sugiyama et al. 1999). These methods predicted the greenfield settlement profile due to tunneling. Then, the building response to tunneling is determined by performing a structural analysis of the building subjected to the predicted settlement profile. In this approach, the following aspects are ignored: (i) structure stiffness, (ii) the interaction between tunneling and adjacent structure, and (iii) 3D nature of the problem. On the other hand,

S. Agaiby and P. Grasso (eds.), *Engineering Challenges for Sustainable Underground Use*, Sustainable Civil Infrastructures, DOI 10.1007/978-3-319-61636-0_8

the use of "Coupled Analysis technique" allows to simulate such complex interaction (Mroueh and Shahrour 2003) which may be achieved through finite element modeling. The finite element approach comprises modelling both the tunnel and the structure in the same model which helps in studying any possible geometrical configuration for the problem.

This paper presents a 3D Finite Element model used to evaluate the performance of a raft foundation in the presence of a tunnel. The paper consists of three parts. The first part validates the tunnel numerical model and its outputs using field data measurements obtained from a case study. The second part addresses the raft modeling and stages of construction. The last part: (i) presenting the settlement of the raft and the induced bending moments during the process of tunnel advancement and (ii) a parametric study to evaluate the effect of related parameters: soil relative density, tunnel diameter, tunnel cover, and horizontal clearance between the raft foundation and the tunnel.

2 Case Study and Model Verification

The selected case study is the Second Heinenoord tunnel in Netherlands based on the work of Moller (2006). This case study includes field measurements for the surface settlement trough during and after the tunneling process in addition to the numerical simulation of the TBM tunneling via the finite element program (PLAXIS)© which is used for the verification and the rest of the parametric study.

The tunnel is a TBM type with slurry shield. The tunnel diameter D = 8.3 m and the soil cover above tunnel crown Z = 12.5 m. The tunnel lining is simulated as a linear elastic concrete material, using shell element with a flexural rigidity EI = 26.78 MN.m^2, normal stiffness EA = 10.5 $\times$ 10^3 MN, weight w = 24 kN/m^2, and Poisson's ratio $\upsilon = 0.15$. The slurry pressure at the face of the tunnel is simulated as an axial pressure = 230 kPa that increases linearly with depth by 15 kPa/m. The grouting at the tail void behind the shield is simulated as a radial pressure = 125 kPa that increases linearly with depth by 15 kPa/m. The TBM shield is simulated as a radial pressure = 125 kPa that increases linearly with depth by 15 kPa/m. The ground-lining gap is simulated by deactivating ground elements with a thickness of 20 cm at the tail of shield. This gap, in the subsequent construction phases, is filled by a hardening grouting material with the same properties of the lining. The TBM is 7.5 m long. The pressurized grouting at the TBM tail is 3.0 m long. The tunnel length is divided into 1.5 m segments to simulate the tunnel advance toward the raft foundation. The total length of the tunnel is 45 m. This length is selected such that the surface settlements reach the steady-state condition.

The tunnel passes through the soil profile shown in Fig. 1. The soil is modeled as a continuum divided into a number of volume elements. The basic soil elements of a 3D finite element mesh are presented by the 10-node tetrahedron elements. Hardening soil model is applied for all soil layers and the various parameters for different soil layers are indicated in Table 1. The ground water table G.W.T is encountered at depth z = 1.5 m from the ground surface.

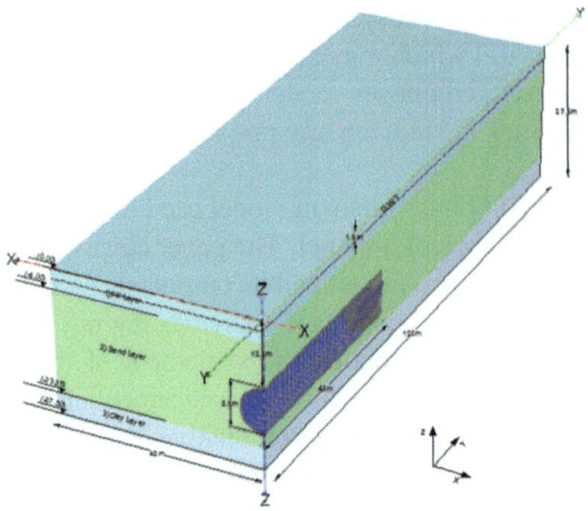

Fig. 1. PLAXIS Three dimensional model configuration for Second Heinenoord tunnel (Moller 2006)

Table 1. Soil properties for the Second Heinenoord tunnel (Moller 2006)

Layer	(1) Fill	(2) Sand	(3) Clay
Saturated unit weight, γ (kN/m^3)	17.2	20	20
Secant stiffness in standard drained triaxial test, E_{50}^{ref} (MPa)	14	35	12
Tangent stiffness for primary oedometer loading, E_{oed}^{ref} (MPa)	14	35	7
Unloading/reloading stiffness, E_{ur}^{ref} (MPa)	42	105	35
Power for stress-level dependency of stiffness, m	0.5	0.5	0.9
Effective cohesion, c′ (kPa)	3	0.01	7
Effective angle of shearing resistance, φ (deg)	27	35	31
Poisson's ratio, υ_{ur}	0.2	0.2	0.2
Earth pressure coefficient at rest, K_o	0.58	0.47	0.55

The generated settlement troughs, in transversal and longitudinal directions, are investigated along the tunnel advance length. Afterwards, they are compared to the settlement trough measured in field during the construction of the Second Heinenoord tunnel. Figure 2 shows a good agreement between the field measurements and the results of the 3D model. The calculated maximum settlement is 25 mm and trough width is about 60 m which match the measured values. The calculated settlement curves are flatter than the measured ones. The steepness of the settlement curve is dependent on the soil constitutive model. In the current study, Hardening soil model is used which provides a flatter curve (Moller 2006).

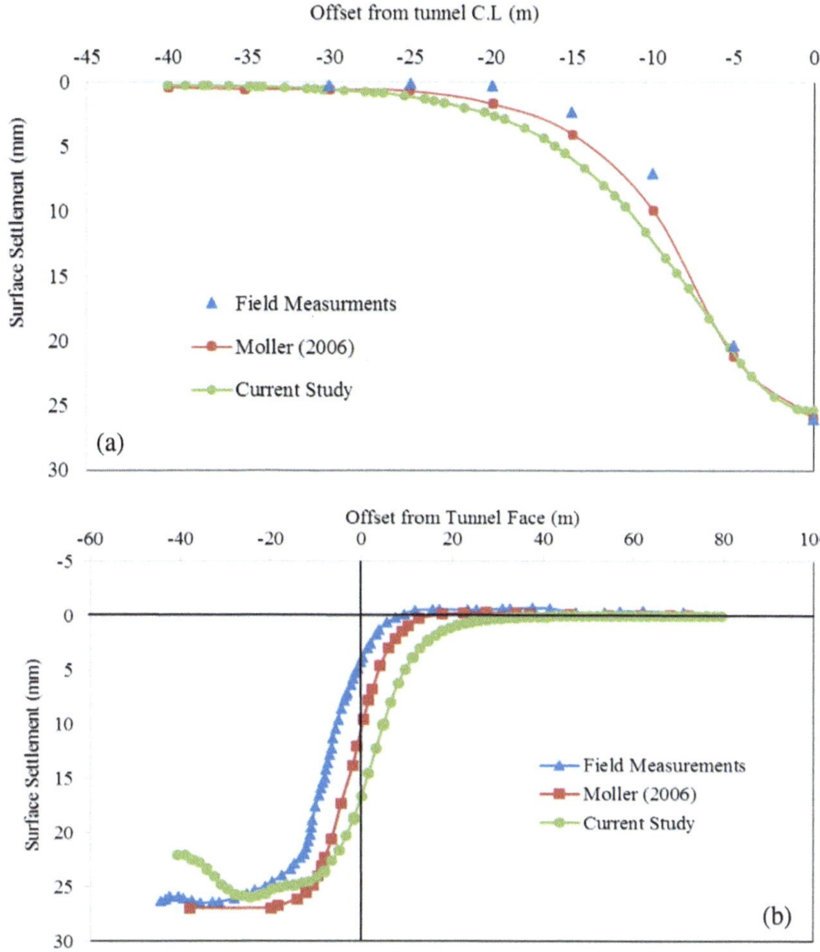

Fig. 2. Surface settlement troughs after verification of the Second Heinenoord tunnel along the: (a) transversal direction, (b) longitudinal direction

3 Raft Modeling and Stages of Construction

The raft characteristics are given by: raft dimensions = 20 m × 20 m, raft thickness = 1.0 m, column's spacing = 4.5 m, loads are applied on the columns such that the average stress underneath the raft = 100 kPa. The raft is simulated as a plate element of a linear elastic concrete material properties: young's modulus E = 2 × 10^7 kN/m^2 and Poisson's ratio $\upsilon = 0.15$.

The model is developed to consider the tunnel stages of construction. First stage, the soil initial stresses are generated using K_o procedure. Second stage, raft foundation element and loads are activated. Third stage, TBM face and shield are activated along the first segment of the tunnel and soil element within this segment is deactivated. Forth stage, as the tunnel advances, TBM face and shield are activated along the next segment and soil element within this segment is deactivated. Meanwhile, the shield in the previous segment is replaced by the tunnel lining. The same procedure of the forth stage is repeated until reaching the end of tunnel length. It should be noted that the model dimensions are selected such that; the borders of the model have no effect neither on raft foundation induced settlements nor on the settlement troughs induced from tunneling (Fig. 3).

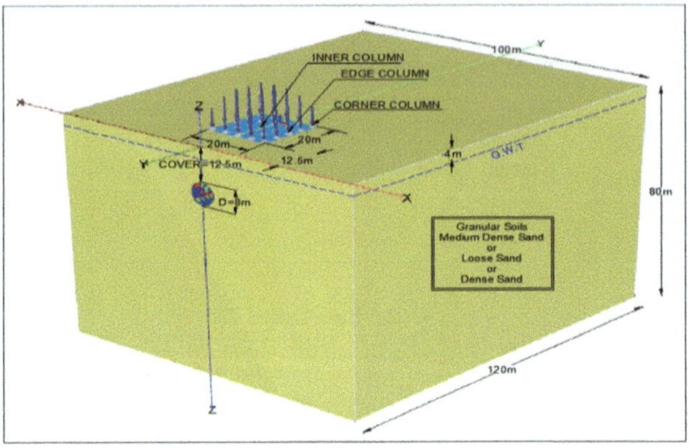

Fig. 3. Proposed configuration of the reference model

4 Effect of Tunnel Advancement and Parametric Study

Figure 4a presents the raft settlement contours before the tunneling process. On the other hand, Fig. 4b shows the raft settlement contours when the tunnel face is beneath the raft centerline. It can be noted that the tunneling process beneath the raft causes the settlement values to increase. In addition, as the tunnel advances, the position of

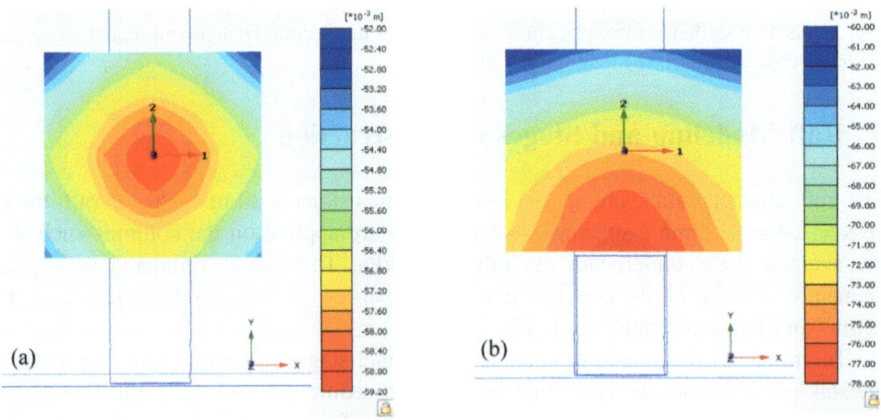

Fig. 4. Raft settlement contours (a) before tunneling process (b) the tunnel face is beneath the raft centerline

maximum settlement changes. Before tunneling process, the maximum settlement value occurs at the center of raft, as the tunnel advances, the maximum settlement value occurs at the edge.

Figure 5a presents the raft bending moment contours before the tunneling process. Whereas, Fig. 5b shows the raft bending moment contours when the tunnel face is beneath the raft centerline. It can be noted that the tunneling process beneath the raft causes the bending moment values to increase. However, the distribution of the bending moment along the raft is unaffected.

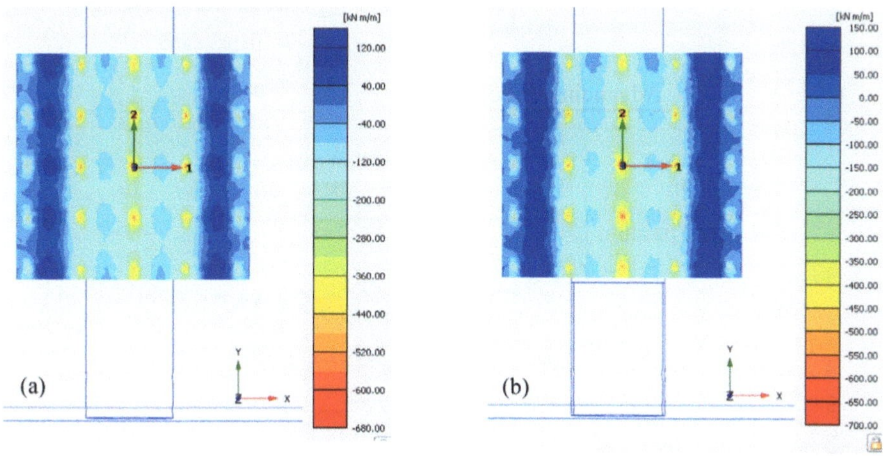

Fig. 5. Raft bending moment contours (a) before tunneling process and (b) the tunnel face is beneath the raft centerline

Figure 6 depicts the evolution of the settlement and the bending moment of the raft foundation during tunneling. Figure 6a shows that the additional maximum settlement of the raft foundation increases during tunneling; it starts when the tunnel face is about 2.7D behind the foundation center line, attains 66–78% of its final value when the tunnel face crosses the raft centerline and then stabilizes when the tunnel face passes the raft foundation centerline by 2.5D. Figure 6b shows the additional differential settlement of the raft foundation. The differential settlement is defined as the difference between the maximum and the minimum raft settlements. The differential settlement value increases during tunneling; attains its maximum value when the tunnel face crosses the raft centerline and then decreases again and stabilizes when the tunnel face passes the raft foundation centerline by 2.5D. The same trend is observed for bending moment (Fig. 6c).

In order to investigate the effect of soil relative density, sand with three different properties presenting loose, medium dense, and dense conditions are proposed in Table 2. It can be noted that the same trend, with respect to tunnel advancement, is achieved for all relative densities (Fig. 6). In addition, an increase in soil relative density leads to a decrease in the tunneling-induced settlements, differential settlements, and bending moments. However, the final value of differential settlement is almost the same regardless the value of soil relative density.

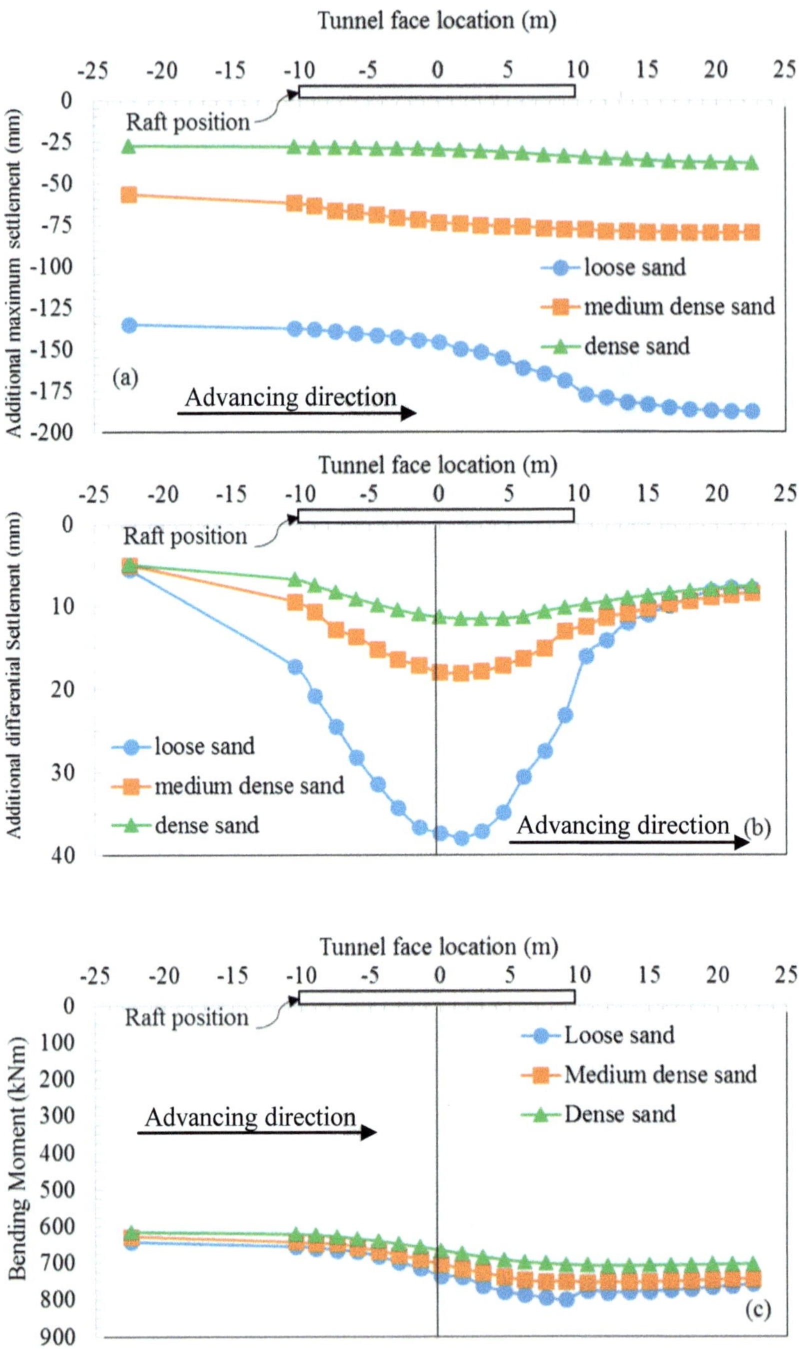

Fig. 6. Raft foundation profiles during tunnel advancement. (a) additional maximum settlement, (b) additional differential settlement, (c) bending moment

Table 2. Proposed different granular soil types of different relative density

Soil type	Loose	Medium dense	Dense
Saturated unit weight, γ (kN/m^3)	17	18	19
Secant stiffness in standard drained triaxial test, E_{50}^{ref} (MPa)	15	35	60
Tangent stiffness for primary oedometer loading, E_{oed}^{ref} (MPa)	15	35	60
Unloading/reloading stiffness, E_{ur}^{ref} (MPa)	45	105	180
Power for stress-level dependency of stiffness, m	0.5	0.5	0.5
Effective cohesion, c′ (kPa)	0.01	0.01	0.01
Effective angle of shearing resistance, φ (deg)	32	35	39
Poisson's ratio, υ_{ur}	0.3	0.25	0.2
Earth pressure coefficient at rest, K_o	0.54	0.47	0.37

Figure 7 investigates the effect of tunnel diameter D on the settlement trough for the same soil type and soil cover. It is noted that as the tunnel diameter increases, the settlements increases. Figure 7 shows that the raft foundation settles 57 mm if no tunnel is constructed. If a 4.0 m tunnel is constructed, settlement increases to 66 mm (16% increase in settlement). If the tunnel diameter increases to 12 m, settlement increases to 96 mm (68% increase in settlement). This trend is noted because as D increases, the loss in soil volume increases. This leads to the increase of the ground settlements. Consequently, the raft settlements are increased as well.

Figure 8 presents the effect of soil cover above tunnel crown Z on the maximum induced values of total and differential settlements. For the same soil type and tunnel diameter, as Z increases the total and differential settlements decrease. However, the maximum settlement does not reach the value of raft total settlement in case of no tunnel. On the other hand, the effect of tunneling on the differential settlement diminishes at Z is equal to 2.5D.

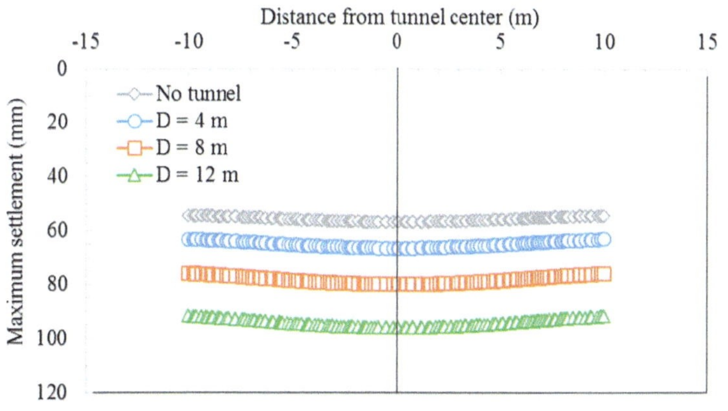

Fig. 7. Effect of tunnel diameter (D) on the settlement trough of raft foundation (medium dense sand, soil cover above tunnel crown = 12.5 m and horizontal clearance = zero)

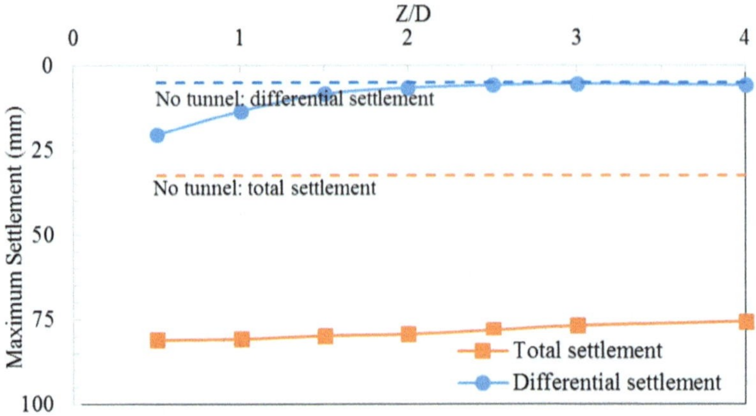

Fig. 8. Effect of soil cover above tunnel crown (Z) on the settlement of the raft foundation (medium dense sand, tunnel diameter = 8 m and horizontal clearance = zero))

Figure 9 depicts the effect of the horizontal clearance between the centerline of the raft foundation and the centerline of the tunnel C on the maximum induced values of total and differential settlements. For the same soil type and tunnel diameter, as C increases the maximum settlements decreases. The effect of tunneling on the settlements vanishes at C is equal to 30 m (C = 2.4Z). However, the maximum value of differential settlement is achieved when the centerline of the tunnel passes underneath the edge of the raft. The maximum value of total settlement is achieved when the centerline of the tunnel passes underneath the middle between the edge and the center of the raft.

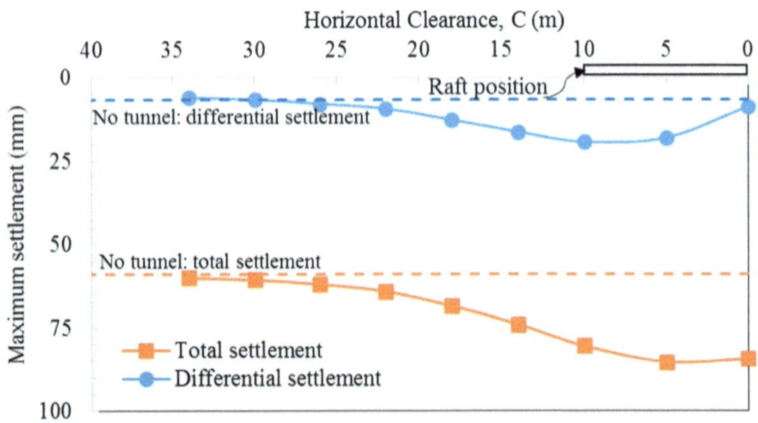

Fig. 9. Effect of horizontal clearance (C) between the centerline of the raft foundation and the centerline of the tunnel on the settlement of the raft foundation (medium dense sand, tunnel diameter = 8 m and soil cover above tunnel crown = 12.5 m)

5 Conclusions

In this study the finite element numerical model succeeded in simulating the complicated TBM shield tunneling method. The model output has been validated through a comparison with the field measurements of the settlement trough recorded during the construction of the Second Heinenoord tunnel in Netherlands (1998). Moreover, the 3D modeling allows to simulate both the tunnel and the raft on the ground surface which helps in studying the complex interaction between the tunnel and the raft with any possible geometrical configuration. In this research, the importance of studying the performance of the raft during the tunneling process is highlighted. The maximum values of differential settlements of the raft occur when the tunnel passes underneath the centerline of the raft foundation. These values are higher than the final values of differential settlements when the tunnel face passes underneath the raft foundation centerline by 2.5D. This behavior could not be depicted by using simple empirical equations or 2D modeling. It is also found that it depends on the sand relative density. For loose sand, the maximum differential settlement is found to be 380% higher than the final differential settlement. For dense sand, the maximum differential settlement is only 50% higher than the final differential settlement. However, the final values of differential settlement are found independent of the sand relative density. It is, also, noted that, during the tunneling process, more than 50% of the additional maximum total settlements and bending moments are attained when the tunnel face crosses the raft centerline and then they were stabilized when the tunnel face passes the raft foundation centerline by a distance equals to 2.5D.

A parametric study is conducted to evaluate the effect of the related parameters; tunnel diameter, tunnel cover, and horizontal clearance between the raft foundation and the tunnel. The raft tunneling-induced settlements are increased as the tunnel diameter increases. The study shows an increase in the total maximum settlement by 16% to 68% when the tunnel diameter is increased from 4.0 m to 12.0 m. This trend may be attributed to the increase in the loss of soil volume as the tunnel diameter increases. It is also noted that the increase in the soil cover above tunnel crown Z leads to the decrease in the maximum induced values of total and differential settlements. The expected raft tunneling-induced differential settlement diminishes at Z is equal to 2.5D. However, raft tunneling-induced maximum settlement are observed down to a depth (Z) equals to 4.0D. Finally, the increase in the horizontal clearance between the center-line of the raft foundation and the centerline of the tunnel leads to a decrease in the raft tunneling-induced settlements. However, the maximum value of differential settlement is recorded at the centerline of the tunnel when the tunnel passes underneath the edge of the raft. Moreover, the maximum value of the total settlement is achieved when the tunnel face passes underneath the mid-distance between the edge and the center of the raft.

References

Maleki, M.: An equivalent beam model for the analysis of tunnel-building interaction. Elsevier Ltd (2011). doi:10.1016/j.tust.2011.02.006

Moller, S.: Tunnel induced settlements and structural forces in lining. Ph.D. thesis, Institute of Geotechnical Engineering, University of Stuttgart (2006)
Mroueh, H., Shahrour, I.: A full 3-D finite element analysis of tunneling–adjacent structures interaction. J. Comput. Geotech. **30**, 245–253 (2003). doi:10.1016/S0266-352X(02)00047-2
O'Reilly, M.P., New, B.M.: Settlements above tunnels in the United Kingdom: their magnitude and prediction. In: Proceedings of the Tunneling 1982, Brighton, pp. 173–181 (1982)
Peck, R.B.: Deep excavation and tunneling in soft ground. In: Proceedings of 7th International Conference on Soil Mechanic Foundation Engineering, Mexico, State-of-the-Art Volume, pp. 225–290 (1969)
Sugiyama, T., Hagiwara, T., Nomoto, T., Nomoto, M., Ano, Y., Mair, R.J., Bolton, M.D., Soga, K.: Observations of ground movements during tunnel construction by slurry shield method at the Docklands light railway lewisham extension – East London. Soils Found. **39**(3), 99–112 (1999)

Three Dimensional Analysis of Diaphragm Walls Supported by Passive Berms

Muhammad M. El-Sherbiny[1(✉)], Rami M. El-Sherbiny[2], and Hussein H. El-Mamlouk[2]

[1] Dar Al-Handasah (Shair and Partners), Cairo, Egypt
mohammed.elsherbiny@dargroup.com,
muhammad.mounir89@gmail.com
[2] Faculty of Engineering, Cairo University, Giza, Egypt
rsherbiny@eng.cu.edu.eg

Abstract. This paper discusses the stability and performance of embedded retaining structures supported by leaving a temporary earth berm on the passive side of the wall in a thick clay deposit characterized by weak soil properties. The objective of this paper is to verify the suitability of using three dimensional finite element software to simulate a bermed excavation. A three-dimensional finite element analysis using PLAXIS 3D software was used to simulate a case history of a monitored retaining wall with a berm in its front side. Analysis and justification of the different parameters used to model the soil and wall is described in this paper. Two different approaches of analysis were applied and compared to account for the stress dependency of soil strength and stiffness. The applicability of the results and adequacy of software were verified by comparing the computed deflections with the monitoring results and a good agreement was realized. Furthermore, the anisotropy of the diaphragm walls was considered to assess its implication on the wall performance during berm removal, as it's deemed to be the most critical stage during construction.

1 Introduction

The technique of leaving earth-berms in place against embedded walls during excavation can provide cost-effective short-term stabilization of the wall before permanent supports are placed (e.g. structural floors). These berms result in an increase of the stabilizing passive pressure acting on the wall due to the increase of size and mass of the passive wedge on the excavation side. Culmann (1866) developed graphical methods to determine the earth pressure due the presence of a berm. Empirical methods evolved in the 1980s and 1990s to evaluate the influence of the berm as a passive thrust on the retaining walls (NAVFAC 1986; Williams and Waite 1993). Potts et al. (1993) performed finite element analysis to evaluate the effectiveness of using berms as a temporary support for retaining walls by using "ICFEP" software. Powrie et al. (1993) used the finite element software CRISP to model the A55 North Wales Coast Road consisting of diaphragm walls and approach embankment in granular soils. Georgiadis and Anagnostopoulos (1998) conducted a 2D finite element analysis using PLAXIS to assess the impact of using an earth berm in front of a cantilever wall embedded in sand.

S. Agaiby and P. Grasso (eds.), *Engineering Challenges for Sustainable Underground Use*, Sustainable Civil Infrastructures, DOI 10.1007/978-3-319-61636-0_9

Morsi (2003) performed broad finite element analyses using Plaxis in a 2-D space, as an approach to quantify the significance of berm geometry as a stabilizing element for retaining walls in cohesion-less soils. However, available literature and design guidelines for berms existing in very soft substrata are limited. The discontinuity of berm during the removal process was most often discounted in research studies. Gourvenec and Powrie (2000) introduced the discontinuous characteristic of a berm during the removal process and its effect on the deflections of walls embedded in thick clay stratum. 3D finite element program CRISP was used to simulate the behavior of this wall using the data obtained from a surveying of the construction of diaphragm walls on the A4/A46 Batheaston bypass in England. The Lias clay soil was modeled as stiff over-consolidated linear elastic – perfectly plastic Mohr Coulomb material. They stated that one of the main drawbacks encountered while modelling the discontinuous berms that exist in a soft strata is specifically the adopted three-dimensional software and generally the processing speed of computers in the time when the study was performed.

This study examines alternative approaches for three dimensional finite element analyses of bermed excavations and provides a comparison of the results with existing analytical solutions. Two methods were presented for modelling the state of stress of the soft clay layers and considering the consolidation of clay. The three-dimensional analyses are intended for further examination of the behavior of the wall with discontinuous berms during berm removal. The anisotropy of the diaphragm wall due to paneled construction becomes of significance when openings exist in a discontinuous wall. Thus, the effect of anisotropy of the diaphragm wall with openings along the berm's body was studied and presented herein.

2 Methodology

Three dimensional finite element software Plaxis 3D (Brinkgreve et al. 2013) was used to model a monitored case history in Taipei City, Taiwan (Liao and Lin 2009). This section presents a description of the case history, numerical modeling approach, material constitutive models, and determination of the soil parameters utilized.

2.1 Description of the Selected Case Study

Liao and Lin (2009) reported field-monitoring data of a diaphragm wall supporting a bermed excavation in Taipei City, Taiwan as part of the Nan-Kang Software Park project. The project consists of a high-rise complex of buildings with three basements. The excavation was in the form of a near-rectangular shape having dimensions of approximately 280 m by 150 m, and a depth of 9.95 m. The excavation was retained using a 0.7 m thick diaphragm wall with a total depth of 19 m below ground surface coupled with a berm on the passive side having a height of 8.25 m and a slope of 1.5H: 1 V. The bermed excavation technique was adopted to merely achieve stability of the wall to support the surrounding soil during excavation of the basement.

The site being located near the alluvium area of the Keelung River, is mainly underlain by stratified silty clay and silty sand layers (Woo and Moh 1990). Medium stiff to stiff silty clay layers extend from the ground level interbedded with loose to medium-dense silty sand layers. Field and laboratory tests at the site consisted of standard penetration testing, unit weight, water content, unconfined compression, and unconsolidated undrained triaxial testing (Liao and Lin 2009); Table 1 presents the interpreted soil parameters. Ground water was encountered at 2 m below ground level.

Table 1. Summary of Subsoil properties (after Liao and Lin 2009)

Parameters	Fill	Silty clay	Silty sand	Silty clay	Silty sand	Silty clay
Depth (m)	0–3.5	3.5–6.5	6.5–10	10–19	19–22	22–28
SPT blow counts, N Value	-	4	6	5	20	7
Water Content, w_n (%)	-	31	25.9	33	24.2	30.2
Liquid Limit, w_L (%)	-	39.1	-	37.6	-	34.1
Plasticity Index, PI (%)	-	16.6	-	15.5	-	13.8
Drained friction angle, ϕ' (^{o})	30	-	30	-	32	-
Undrained Shear Strength, S_u (kPa)	-	24-34	-	35-45	-	50-65

2.2 Numerical Modeling

Due to the large footprint of the construction site, a slice of the bermed excavation was selected expressing the berm as extending continuously in out-of-plane direction. A three dimensional finite element mesh was built for this particular case study. The length of the mesh was 12.5 m in the x-direction, the width was 55 m in the y-direction and the height was 28 m in the z-direction as shown in Fig. (1). The horizontal boundary was taken at a distance 2.5 times the width of the berm so that the boundary effect would be minimal. The change in pore water pressure with depth from the initial hydrostatic case due to dewatering works was accounted for during the progress of excavation.

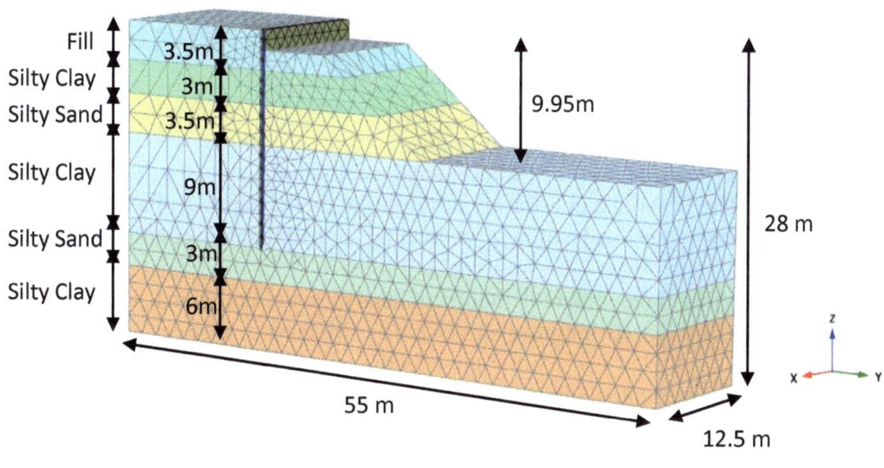

Fig. 1. The conducted 3D model on Plaxis for the selected case study

Liao and Lin (2009) didn't discuss the times at which the deformation readings were monitored. Moreover, they didn't specify the duration for the excavation of berm. This posed a challenge to the analysis and reasonable estimates were therefore considered. Gourvenec and Powrie (2000) suggested that the minimum duration for an excavation of a bermed temporary structure is 1 month. On such basis, it was assumed that the bermed excavation would be standing for 30 days. The simulations were extended to 150 days to further investigate the behavior as it can be considered as an upper limit for a temporary structure.

The numerical model consisted of five phases as follows:

Initial phase: Initial state of stresses for soil before the commencement of any works.
Phase 1: Installation of the diaphragms walls.
Phase 2: Dewatering below the excavation level.
Phase 3: Excavation to the target depth by forming the berm body.
Phase 4: Elapsed time of 30 days after excavation of berm.
Phase 5: Elapsed time of 150 days after excavation of berm.

2.3 Structural Elements

An isotropic plate of a linear elastic model was adopted for numerically modeling the wall element. The wall thickness (d), unit weight (γ), wall modulus in both directions (E_1 and E_2), Poisson ratio (ν), and shear modulus (G_{12}, G_{13} and G_{23}) used in modeling the wall are presented below in Table 2. The horizontal wall modulus was varied in the model with a discontinuous berm to study the effect of wall anisotropy.

Table 2. Isotropic elastic parameters used to model Diaphragm wall

Element	d (m)	γ (kN/m^3)	E_1 (kPa)	E_2 (kPa)	ν	G_{12} (kPa)	G_{13} (kPa)	G_{23} (Kpa)
Plate	0.7	25	2e7	2e7	0.15	8.7e6	8.7e6	8.7e6

2.4 Soil Models and Properties

Soil layers were modeled as three-dimensional continuum using the hardening soil model (HSM). The HSM is a non-linear elasto-plastic model capable of accounting for stress dependency of the soil stiffness in loading and unloading conditions Schanz (1998).

Plaxis 3D offers three different modes of analyses for undrained soil behavior, methods (A, B, and C) (Brinkgreve et al. 2013). Method A enables modeling the undrained behavior of clay considering the effective stress parameters for strength and stiffness. The undrained calculations are performed in the form of effective stress analysis considering pore pressure generation and adjusting the stiffness to account for the undrained behavior. Thus, the undrained strength and stiffness of the clay is an output. The resulting undrained shear strength is therefore crosschecked with measurements.

A consolidation analysis can be performed with the undrained calculation to account for time dependent shear strength and deformations.

Method B enables performing the undrained calculations considering effective stiffness parameters and undrained shear strength. However, consolidation analyses cannot be performed using this method. Method (C) is available only for the Mohr-Coulomb and liner elastic soil models, and was therefore not considered. For the current research, Methods (A) and (B) were adopted and compared.

No direct measurements were available on the deformation characteristics of the different soil layers. The stiffness of cohesive and cohesion-less soils were therefore estimated based on available field and laboratory test results. For cohesive soils, Duncan and Buchignani (1976) introduced a relationship between the undrained modulus of elasticity normalized by the undrained shear strength (s_u), the plasticity index of soil and the over consolidation ratio (OCR). The OCR ratio was calculated using the following relationship (Kulhawy and Mayne 1990):

$$s_u/\sigma'_{vo} = (0.23 \pm 0.04)\mathrm{OCR}^{0.8} \tag{1}$$

The effective-stress angle of shearing resistance for clays was estimated based on correlations. The encountered cohesive soils were found to be mainly in a normally consolidated state of stresses. Sorensen and Okkels (2013) discussed correlations between the plasticity indices for primarily normally consolidated clays versus the corresponding effective angle of friction collected from triaxial test results. Accordingly, a direct correlation can be established between the effective shear strength and plasticity index of cohesive soils as follows,

The effective stress friction angle (ϕ') of the clay was estimated using the correlation proposed by Sorensen and Okkels (2013) for primarily normally consolidated clays as a function of plasticity index (I_P):

$$\phi' = 43 - 10 * \log I_p \tag{2}$$

Consolidated undrained triaxial (CUTX) compression tests were simulated using Plaxis for the different layers of silty clay to compare the measured undrained shear strength with the values obtained using effective stress parameters. The tests were simulated assuming anisotropic consolidation (k_o condition) in which k_O is considered 0.5 (i.e. $\sigma'_3 = 0.5\,\sigma'_1$). Figure (2) depicts the stress-strain curves obtained from simulating the silty clay layers encountered in the case study. Results of simulated anisotropic CUTX tests show an overall agreement with the field test results except for occasional minor differences.

For cohesion-less soils, the drained young's modulus was deduced from the SPT blow counts using the relationship introduced by Callanan and Kulhawy (1985). The values of angle of shearing resistance presumed by Liao and Lin (2009) for the cohesion-less soils were considered reasonable and were used in the analyses. Table 3 represents the adopted parameters for the different soil layers included in the numerical model.

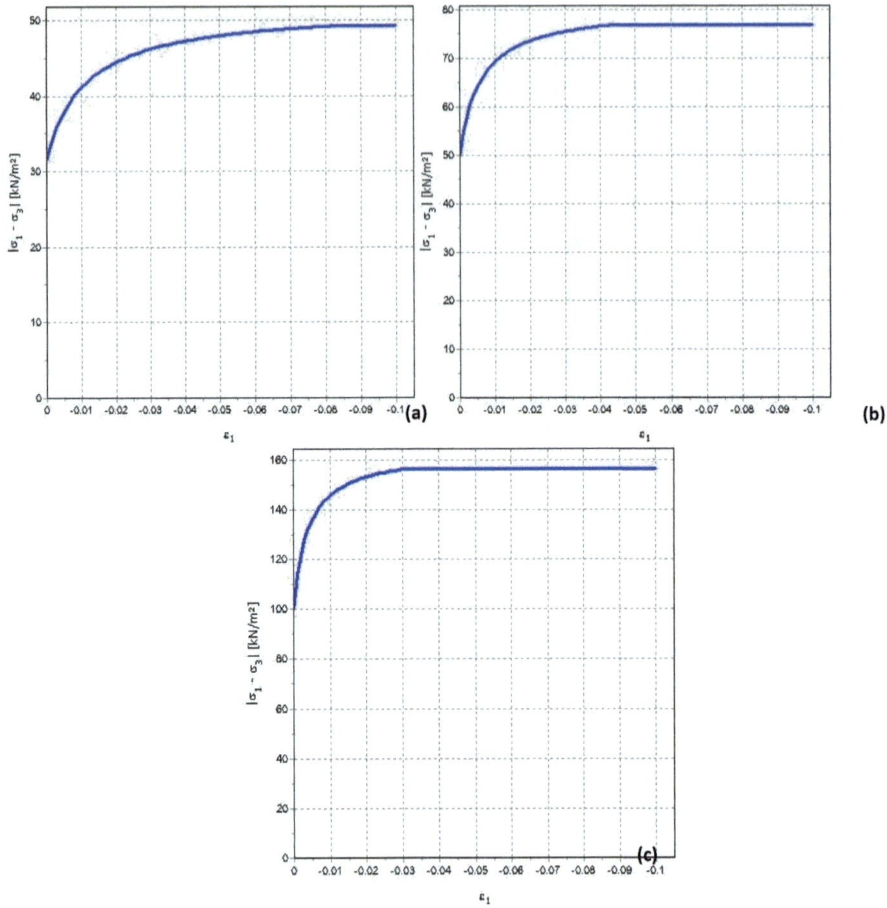

Fig. 2. Simulation of anisotropic CUTX on silty clay layers at depths of: (a) 3.5 m–6.5 m, (b) 10 m–19 m & (c) 22 m–28 m

Table 3. Summarized material properties adopted for the soil verification model

Parameters	Fill	Silty clay	Silty sand	Silty clay	Silty sand	Silty clay
Depth (m)	0–3.5	3.5–6.5	6.5–10	10–19	19–22	22–28
Model	HSM	HSM	HSM	HSM	HSM	HSM
Drainage type	Drained	Undrained	Drained	Undrained	Drained	Undrained
γ_{sat} (kN/m^3)	18.6	18.6	18.6	18.6	18.6	18.6
E_{50}^{ref} (kPa)	10000	8000	20000	15000	50000	25000
E_{oed}^{ref} (kPa)	10000	8000	20000	15000	50000	25000
E_{ur}^{ref} (kPa)	30000	24000	60000	45000	150000	75000
C' (kPa)	1	1	1	1	1	1
Ø' (degrees)	30	31	30	31	32	32
R_{inter}	0.7	1	0.7	0.9	0.7	0.7

3 Results

In order to verify the suitability of using the proposed three-dimensional finite element model to simulate a bermed excavation, computed deflections from the numerical model are compared with the monitored deflections obtained from the field inclinometers at each side of the excavation (Fig. 3). The analyses were established using Method (A) (i.e. considering the drained soil strength and stiffness parameters in an undrained analysis). The computed lateral deformations of the wall at 30 days after the commencement of constructions show a good match with the monitored values along the entire depth of the wall (i.e. at depth −12.5 m, deflections at 30 days was 3.4 cm which lies in between the monitored deflections of the north wall 2.5 cm and the south wall 3.8 cm)

The Method (A) analysis was compared with a Method (B) for modeling the undrained behavior of the silty clay layers considering the undrained strength parameters obtained from field tests. The Method (B) analysis was established using a plastic

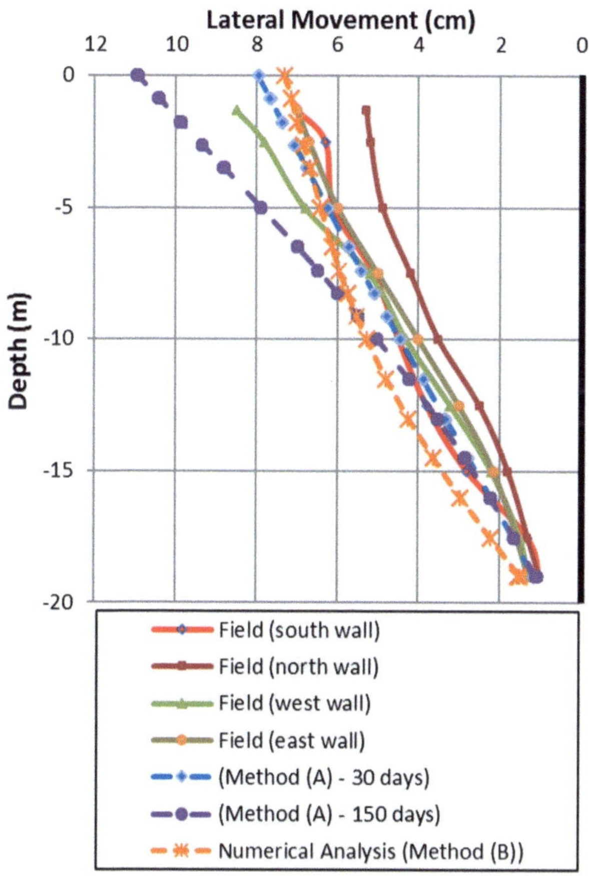

Fig. 3. Comparison of lateral deformations from finite element analyses and field measurements

calculations phase rather than a consolidation phase. The computed deformations using Method (B) analysis showed higher deformations than method (A) at larger depths and slightly lower deformations at the top. The overall behavior of the wall was better captured by the Method (A) analysis when compared to the field measurements. However, the computed vertical settlement behind the wall from Method (B) analysis showed a good agreement with the corresponding settlements from Method (A) analysis at 30 days.

Extending Method (A) analysis beyond 30 days resulted in increased lateral deformations at the top of the wall, beyond the values measured in the field. Most of the vertical deformations behind the wall from the consolidation analysis of Method (A) occurred in the earlier periods of the construction within 150 days after the commencement of construction Fig. 4. No significant increase occurred in the vertical settlements behind the wall until reaching a 90% consolidation.

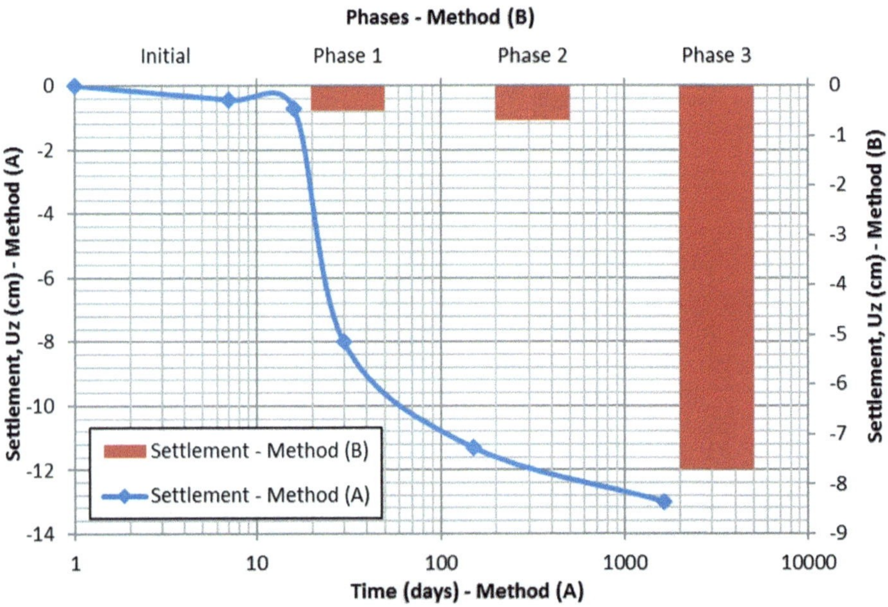

Fig. 4. Vertical deformations behind the wall from finite element analyses

According to the results of the simulations, Method (A) analysis is considered a more adequate approach to simulate the temporary bermed excavation in case it exists in an area of thick clay stratum characterized by weak properties. On such basis, a broad sensitivity analysis can be performed to develop a better understanding of the factors affecting bermed excavation behavior and behavior during berm removal.

4 Comparison with Analytical Methods

4.1 Effect of Wall Anisotropy

The stiffness of the diaphragm walls was considered isotropic in the verification model.

Finno et al. (2006), Lee et al. (1998) and Zdravkovic et al. (2005) introduced the variance of wall properties in vertical and horizontal directions to simulate anisotropy. Limited literature addressed the effect of wall's properties variance with the existence of a bermed excavation. Diaphragm walls are constructed in the form of discrete panels placed beside one another. The wall stiffness should be reduced in the lateral direction where the lateral reinforcement isn't continuous. Three cases of lateral stiffness were chosen for modeling the wall, together with the effect of using capping beams. These are identified as follows:

Isotropic Model (Default model): The ratio of the lateral stiffness to the vertical stiffness is taken as one (i.e. $E_2/E_1 = 1$).

First Model: The ratio of the lateral stiffness to the vertical stiffness is taken as 1/1000 (i.e. $E_2/E_1 = 1/1000$).

Table 4. Properties of the selected capping beam

Dimensions	Width (m)	Depth (m)	I_2 (m^4)	I_3 (m^4)
0.7 × 1	0.7	1	0.02858	0.05833

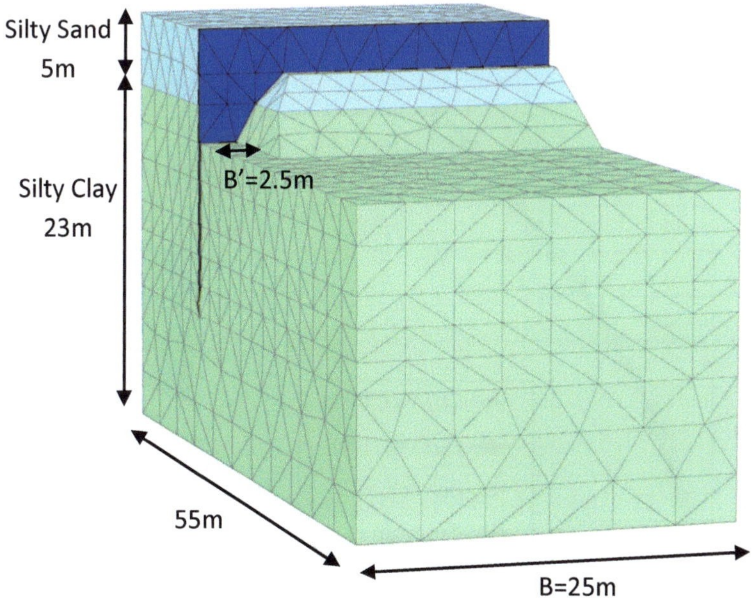

Fig. 5. The simplified 3D model on Plaxis for the case of discontinuous berm

Second Model: The ratio of the lateral stiffness to the vertical stiffness is taken as 1/100 (i.e. $E_2/E_1 = 1/100$).
Capping Beam Model: properties of capping beam are presented in Table 4.

In order to evaluate the degree of influence of the wall anisotropy, the deflections were investigated at the crest of the diaphragm wall while the berm exists in an intact condition for a simplified soil model Fig. 5. Table 5 represents the soil properties adopted for the simplified soil model. A comparison of the lateral deformations from

Table 5. Summarized material properties adopted for the simplified soil model

		Materials	
		Silty sand	Silty clay
Parameters	Depth (m)	0–5	5–28
	Model	HSM	HSM
	Drainage type	Drained	Undrained
	γ_{sat} (kN/m3)	18.6	18.6
	E_{50}^{ref} (kPa)	45000	10000
	E_{oed}^{ref} (kPa)	45000	10000
	E_{ur}^{ref} (kPa)	135000	30000
	C' (kPa)	1	1
	Ø' (degrees)	34	32
	R_{inter}	0.8	0.9

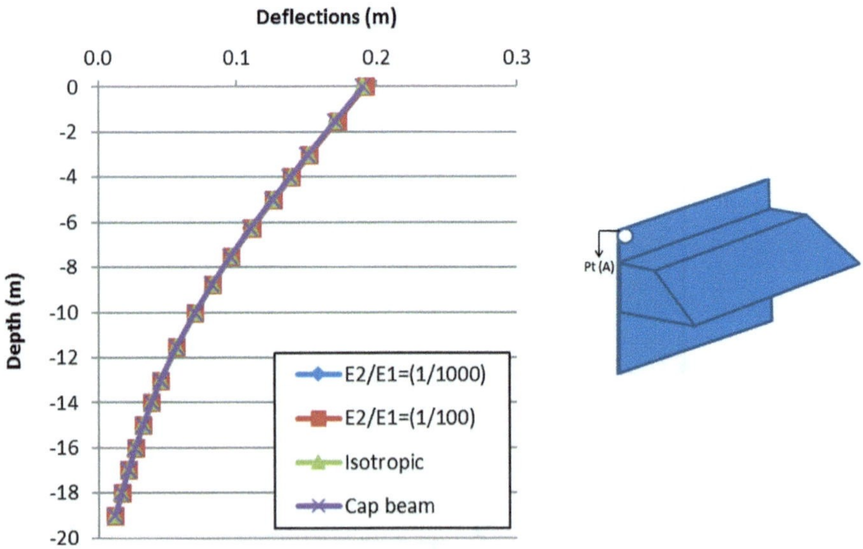

Fig. 6. Deflection's profiles for different wall models at Pt. (A) for a berm in an intact condition following Method (A) analysis (opening width; 2B' = 0 m and opening spacing; 2B = 50 m)

the different cases in presence of an intact berm Fig. 6 shows no significant difference. This is due to the presence of the berm's body in an intact condition leading to plane strain conditions with no effect for the out of plane stiffness of the wall.

For a better evaluation of the anisotropic phenomenon of the diaphragm walls, openings were introduced through the berms body with an opening's width (2B') = 5 m and a spacing between openings (2B) = 50 m (i.e. following the theory of symmetry, half opening width B' = 2.5 m and half spacing between openings B = 25 m). The effect of wall anisotropy resulted in difference in deflection reaching around 13% at the maximum deflection Fig. 7. This effect could be a result of the sequence of berms providing intermittent support that leads to arching in the soil behind the wall. Yet, this point needs further investigation to highlight the relation between discontinuous berms and the arching effect for the soil behind the wall.

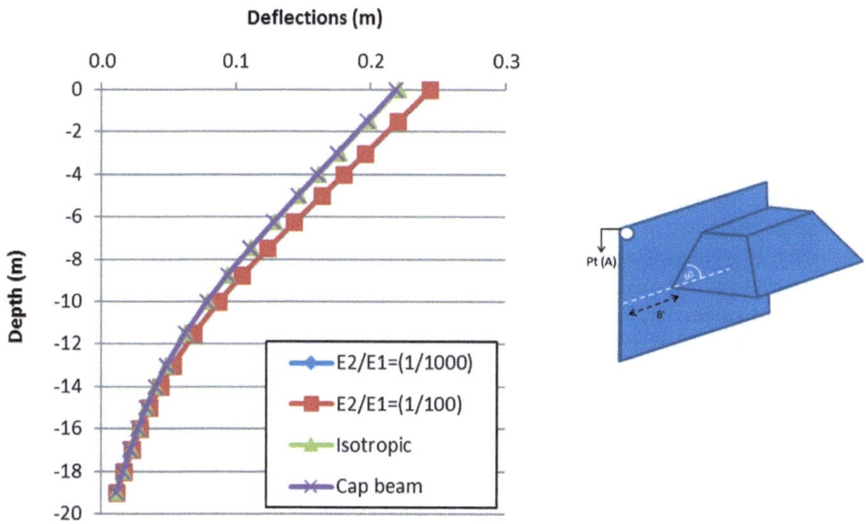

Fig. 7. Deflection's profiles for different wall models at the center of opening Pt. (A) following method (A) analysis (opening width; 2B' = 5 m and opening spacing; 2B = 50 m)

5 Conclusions

This study is mainly concerned with numerically simulating the behavior of berm as temporary stabilizing structure for embedded walls, especially when it exits in a thick clay substratum characterized by weak properties. Three-dimensional finite element analyses were utilized to simulate a case history of a retaining wall with stabilizing berm in the passive side. Soil properties were selected based on available field and laboratory tests while the missing data were estimated using well established correlations and justified by simulating the results of triaxial tests. Two modeling approaches were considered, Method (A) where the soil stiffness and strength are defined in terms of effective stress, and Method (B) where the soil stiffness was defined in terms of effective stress while the shear strength was defined in terms of total stress.

The results of finite element simulations using Method (A) within 30 days of construction were in good agreement with the measured lateral deformations of the wall. However, Method (B) analysis overestimated the lateral deformation at large depths and under estimated the deformations at the top. Lateral deformations of the wall and vertical deformations behind the wall tended to increase with time, but the increase faded after 150 days of construction. Anisotropy of the diaphragm wall stiffness was considered to simulate the effect of the discrete paneling of the wall for a discontinuous berm in the removal process. Difference in deformations about 13% was observed between modeling the wall in isotropic and anisotropic conditions. This reflects that the anisotropy of the wall should be considered especially during removal process of the berm.

References

Brinkgreve, R.B.J., Engin, E., Swolfs, W.M.:. Plaxis 3D 2013. Plaxis bv, Delft, Netherlands (2013)

Callanan, J.F., Kulhawy, F.H.: Evaluation of procedures for predicting foundation uplift movements. Report EL-4107, Electric Power Research Inc., Palo Alto, p. 124 (1985)

Duncan, J.M., Buchignani, A.L.: An engineering manual for settlement studies. University of California, Berkeley, p. 94 (1976)

El-Sherbiny, M.: Numerical analysis of embedded retaining walls supported by discontinuous unimproved and improved berms. M.Sc. thesis, Cairo University, Giza, Egypt (2016)

Finno, R.J., Blackburn, T., Roboski, J.F.: Three-dimensional effects for supported excavations in clay. J. Geotech. Geoenviron. Eng. **133**(1), 30–36 (2006)

Georgiadis, M., Anagnostopoulos, C.: Effect of berms on sheet-pile wall behaviour. Geotechnique **48**(4), 569–574 (1998)

Gourvenec, S., Powrie, W.: Three-dimensional finite element analyses of embedded retaining walls supported by discontinuous earth berms. Can. Geotech. J. **37**(5), 1062–1077 (2000)

Kulhawy, F.H., Mayne, P.W.: Manual on Estimating Soil Properties for Foundation Design. Cornell University, Ithaca (1990)

Lee, F.H., Yong, K.Y., Quan, K., Chee, K.T.: Effects of corner in strutted excavations: field monitoring and case histories. J. Geotech. Geoenviron. Eng. **124**, 339–349 (1998)

Liao, H.J., Lin, C.C.: Case studies on berm excavation in Taipei silty soil. Can. Geotech. J. **46**(8), 889–902 (2009)

Morsi, Y.G.: Berms for stabilizing earth retaining structures. M.Sc. thesis, Cairo (2003)

Naval Facilities Engineering Command: Design manual NAVFAC DM-7.02, Foundation and earth structures. U.S. Department of navy, Washington, DC (1986)

Potts, D.M., Addenbrooke, T.I., Day, R.A.: The use of soil berms for temporary support of retaining walls. In: Proceedings of the Conference Retaining Structures, Cambridge, pp. 440–447 (1993)

Powrie, W., Davie, J.N., Britto, A.M.: A cantilever retaining wall supported by a berm during temporary works activities. In: Proceedings of the Conference Retaining Structures, Cambridge, pp. 418–428 (1993)

Schanz, T.: Zur Modellierung des Mechanichen Verhaltens von Reibungsmaterialen. Habilitation, Stuttgart Universitat (1998)

Sorensen, K.K., Okkels, N.: Correlation between drained shear strength and plasticity index of undisturbed overconsolidated clays. In: Proceedings of the 18th International Conference on Soil Mechanics and Geotechnical Engineering, Paris, pp. 1–6 (2013)

Williams, B.P., Waite, D.: The design and construction of sheet-piled cofferdams. CIRIA Special Publication 95, Construction Industry Research and Information Association, London (1993)
Woo, S.M., Moh, Z.C.: Geotechnical characteristics of soils in the Taipei Basin. In: Proceedings of the 10th Southeast Asian Geotechnical Conference, Taipei, vol. 2, pp. 51–65 (1990)
Zdravkovic, L., Potts, D.M., St John, H.D.: Modeling of a 3D excavation in finite element analysis. Geotechnique, **55**(7) (2005)

Behaviour of Urban Metro Twin Tunnels Under Earthquake Loads

R.B. Jishnu(✉) and Ramanathan Ayothiraman

Department of Civil Engineering, IIT Delhi, New Delhi 110016, India
jishnurb@gmail.com, araman@civil.iitd.ac.in

Abstract. In Indian cities like Delhi, where seismic demand is on a higher side, it will be prudent enough to perform seismic analysis of underground structures like tunnels, which in this study, is the primary focus. The geometry of Delhi Metro and specific soil conditions experienced in Delhi are considered for the analysis. The seismic analysis was carried out using pseudo-static approach, considering site conditions. In order to appreciate the effect of earthquake loads, analyses were carried out for both static loads and seismic loads, separately. The response of tunnel liner in terms of moments, thrusts and shear are predicted under both loads. However, the effect of seismic loads on these response parameters is quantified in terms of additional moments, thrusts and shear respectively. Also, emphasis is given to quantify surface displacements due to tunneling under earthquake loads, in addition to static loads. It is found that the response of twin tunnels is affected by earthquake loading, which may be incorporated in the design.

1 Introduction

Due to high rate of urbanization and increase in civilian transportation demands, metro rail systems are becoming quite common. As utilizable space on the surface is comparatively less, especially in unplanned cities, construction of underground metro structures in the form of *NATM* (New Austrian Tunneling Method)/Bored tunnels is on the rise. In Indian cities like Delhi, where seismic demands are high it is imperative to perform a seismic analysis for these underground facilities to ensure the strength and serviceability of its structural components. This is despite the general notion that underground structures are deemed safe against earthquakes and hence, such demands could be ignored. For underground structures inertia of surrounding soil is large relative to inertia of the structure. This effectively means that the response of tunnel is dominated by the surrounding ground rather than the tunnel lining properties, which being an absolute contrast to surface structures. The importance of seismic analysis for underground structures constructed in Delhi, where earthquake zonal factors are high (IS 1893 2002, Table 2, Clause 6.4.2), cannot be ignored. Hence such analyses are being performed to come up with significant observations.

For this analysis, typical geotechnical profile of Delhi is considered. Here strength parameters of the soil are governed by Mohr-Coulomb criterion. Underground metro tunnels which normally occur in twin configuration ("up" line and "down" line), are assumed to be at different cover depths i.e. $1D_{ex}$, $1.5D_{ex}$, $2D_{ex}$, $2.5D_{ex}$ where 'D_{ex}' represents excavated diameter. Cover depths less than $1D_{ex}$ are ignored in this analysis

S. Agaiby and P. Grasso (eds.), *Engineering Challenges for Sustainable Underground Use*, Sustainable Civil Infrastructures, DOI 10.1007/978-3-319-61636-0_10

as normally ground improvement methods are adopted in those cases due to tunnel stability reasons. Staged analysis is performed for tunnel excavation sequence incorporating tunnel "lag" for the second tunnel. For pseudo-static analysis, zonal factors suggested in *IS 1893* (2002) are used. Structural response of the tunnel liner at different cover depths are monitored for these zonal parameters along with ground displacements. Dry conditions are assumed throughout the analysis and hence dynamic pore pressure build up could be ignored.

1.1 Geotechnical and Tunnel Lining Parameters

For the numerical analysis, a typical geotechnical profile of Delhi (Hanumantharao and Ramana 2008) corresponding to silty sands is considered as majority of Delhi is characterized by such deposits (Hanumantharao and Ramana 2008). Various parameters for this strength criterion under static condition are ground modulus (*E*), Poisson's ratio (v), cohesion (c), friction angle (ϕ) and soil weight density (γ). The shear strength parameters at various depths, for typical boreholes under consideration have been derived from Unconsolidated Un-drained (*UU*) Tests. Ground Modulus (*E*) have been computed from field *SPT* measurements using correlations provided for Silty Sands (Bowles, 1988). Here for the considered soil profile (typically Silty Sand, *SM*) appreciable variation is not observed for soil weight density (γ), friction angle (ϕ), and Poisson's ratio (v). Hence in this study, γ = 19 kN/m^3, ϕ = 17° and ν = 0.33 are considered constant. The variation of ground modulus (*E*) and cohesion(c) along with depth is shown in Fig. 1. For the staged excavation of these tunnels (with suitable lag) above parameters were considered for the numerical analysis.

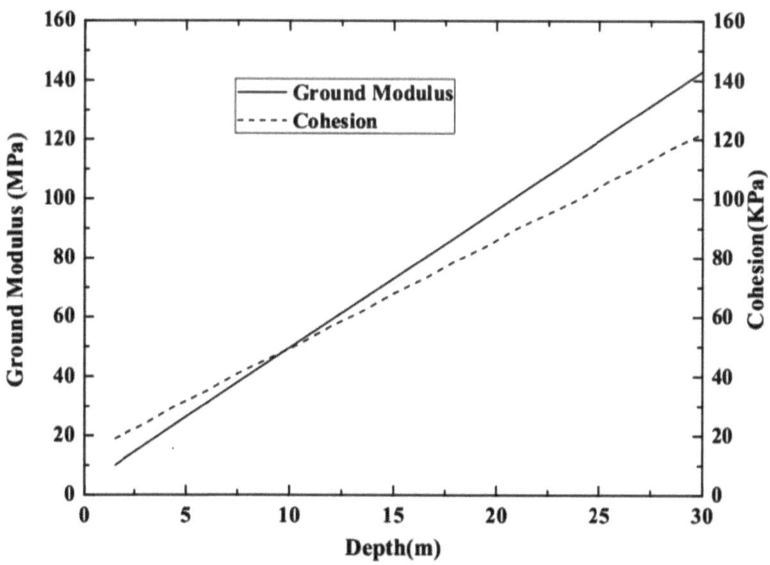

Fig. 1. Variation of ground modulus and cohesion with depth for the considered geotechnical profile

For the tunnel lining, material is considered to be linear elastic. Twin tunneling operation is performed in sequence with appropriate lag distance. For metro tunnels in Delhi an internal diameter of 5.8 m is used with liner thickness of 0.275 m. These tunnels have a typical 5 + 1 configuration (i.e. 5 main segments and a key segment). During plane strain analysis effect of longitudinal joints are considered in the form of reduction factors (Blom 2009) based on number of joints, contact area of longitudinal joint and tunnel radius. By this method an equivalent continuous liner with reduced bending stiffness incorporating presence of joints could be obtained.

1.2 Static Analysis by Staged Excavation for Twin Tunnels

Numerical analysis is performed for twin tunnel excavation using *FLAC2D*. For this analysis the following stages are performed: (1) Geostatic condition; (2) In-situ stress relaxation of first tunnel to simulate the three dimensional arching effects of tunneling; (3) Stabilization of excavation using segmental lining and allowing 100% stresses to relax; (4) In-situ stress relaxation of second tunnel to simulate the arching effects; (5) Stabilization of excavation using segmental lining for the second tunnel and allowing 100% stresses to relax; and (6) Pseudo-static analysis using seismic zonal parameters.

Construction Stage Analysis of Twin Tunnels. First the geostatic condition or in-situ stress condition is established for the soil profile using 'K_0' procedure. The resultant displacements are nullified before the second stage. For Stage 2 (relaxation of first tunnel), appropriate relaxation factor inside the tunnel boundary has to be calibrated. In urban scenario surface displacements induced by the tunnel plays a significant role. Hence stress relaxation inside the tunnel is controlled in order to achieve a targeted volume loss value. For metro tunnels this value could be around 0.5% to 1% of excavated volume. A conservative volume loss value of 0.5% is assumed for this analysis. A stress relaxation of 30% inside the tunnel periphery is found to be adequate (for different cover depths $1D_{ex}$, $1.5D_{ex}$, $2Dex$, $2.5D_{ex}$) to achieve this targeted value of volume loss. Once this relaxation is applied for the first tunnel, surface settlements in the form of Gaussian profiles are obtained and this is compared with classical (Peck 1969) curves. One such comparison at a typical cover depth ($1.5D_{ex}$) is shown in Fig. 2. The ratio of the distance between tunnel centre and point of inflexion (i) with the tunnel depth (z, up to springing line) is termed as 'k' parameter. This 'k' parameter from the numerical analysis is compared with the values reported by Lake et al. (1992). It is found that for all the cover depths, the 'k' parameter varies from 0.58–0.6 which is in agreement with the values for similar soil deposits (0.6), reported by Lake et al. (1992).

According to Peck (1969), with the trough width being proportional to the depth (with 'k' parameter being proportionality constant), it is expected that with the increase in cover depth, there could be an increase in trough width with a reduced settlement for a constant volume loss. This trend of decreasing settlement and increasing trough width with an increase in tunnel depth is numerically verified and it is shown in Fig. 3.

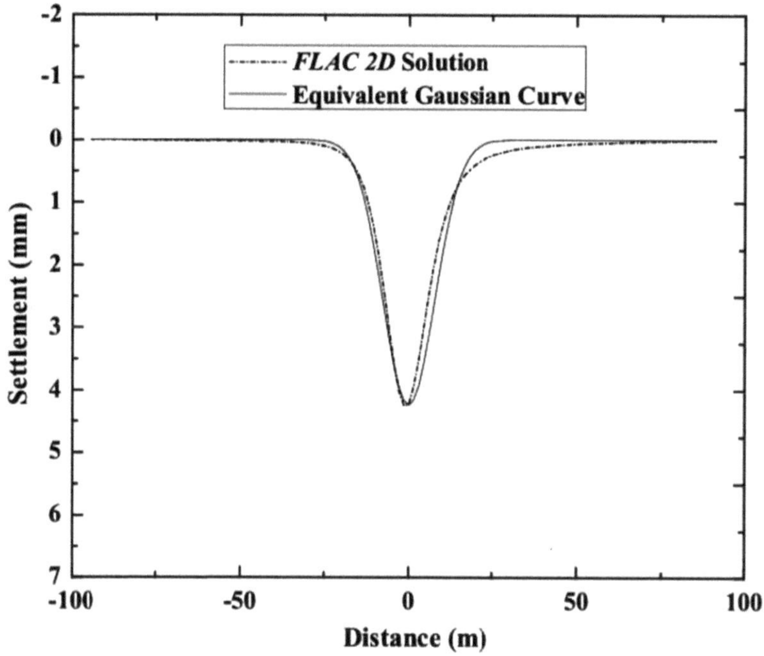

Fig. 2. Comparison of subsidence profiles (*FLAC 2D* Vs Peck 1969 - k = 0.6)

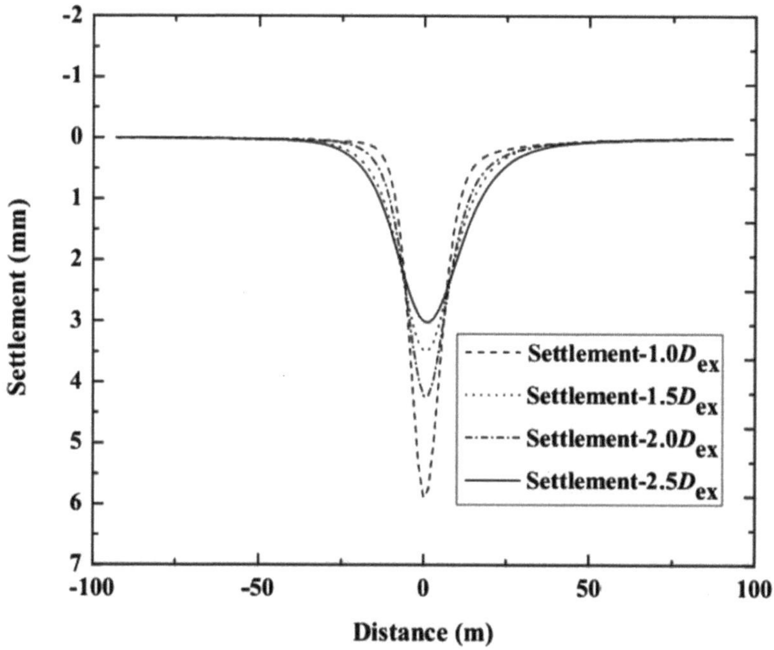

Fig. 3. Comparison of subsidence profile (FLAC 2D) for a tunnel at various cover depths

To confirm whether liner elements works in conjunction with the ground mass around, the tunnel liner forces are compared with classical equations given in literature (Bobet 2001). This method is adopted as it gives distribution of moments at any salient point around the tunnel periphery. Since these methods could not cater stress relaxation, the validation is performed for such a condition. A typical result for validation of liner forces, in these cases moments, obtained from numerical solution is compared with classical equation. The comparison is shown in Fig. 4.

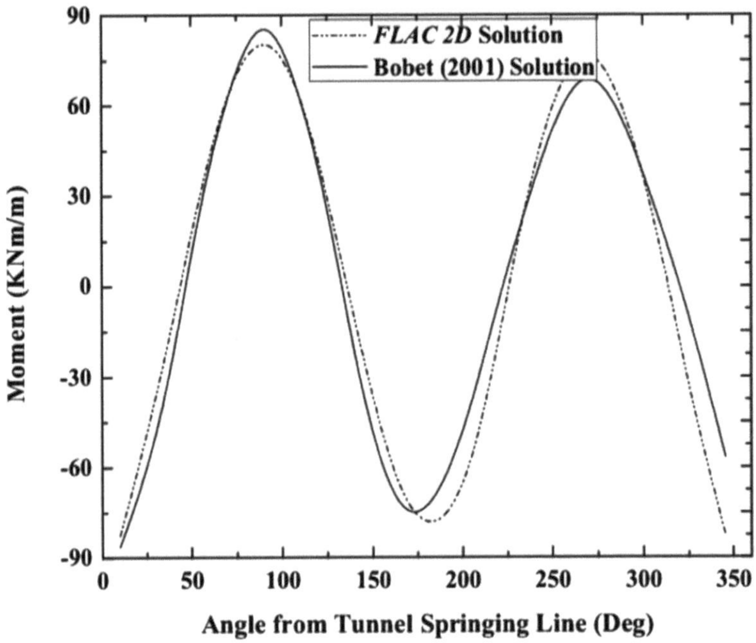

Fig. 4. Comparison of tunnel liner moments (*FLAC 2D* Vs Bobet 2001)

As already mentioned, the second tunnel is excavated after the construction of first tunnel liner with an appropriate lag. Due to the construction of second tunnel, the surface settlement profiles will be modified with more settlements occurring on the latter constructed tunnel (Chapman et al. 2007). The development of asymmetric settlement curves due to twin tunneling could be observed in this analysis as well. For validation of such curves, super position principle is used. It could be observed that these super position curves are relatively in good agreement with numerical solution (Fig. 5). At this point it should be mentioned that as per literature (Kim et al. 1996; Chapman et al. 2007) superposition of single tunnel settlements could not be used for twin tunnel settlement deduction. Hence it is always advisable to perform a surface settlement analysis for twin tunnels using numerical methods. Also the decrease of maximum surface settlements for a constant volume loss could be observed even for twin tunnels, for an increase in cover depth (Fig. 6).

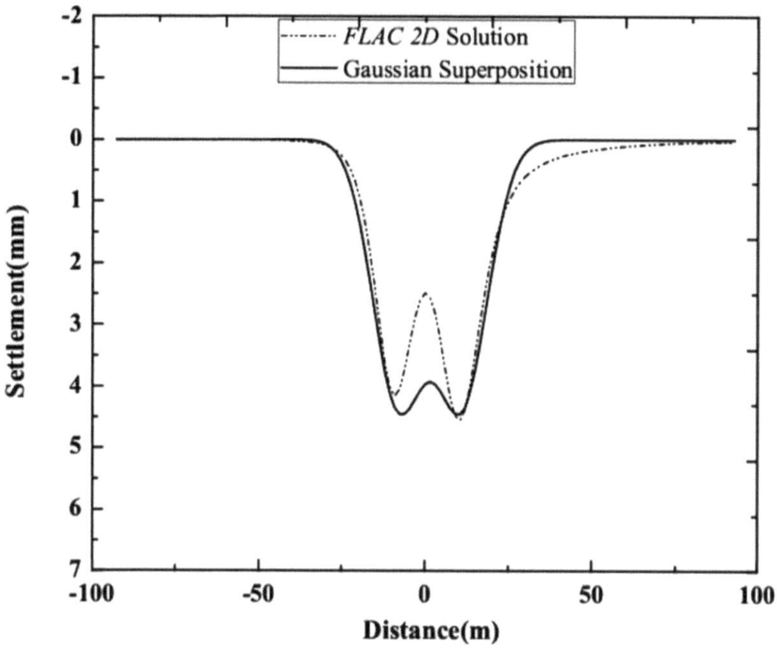

Fig. 5. Comparison of subsidence profile for twin tunnels (*FLAC 2D* Vs superposed Gaussian profiles)

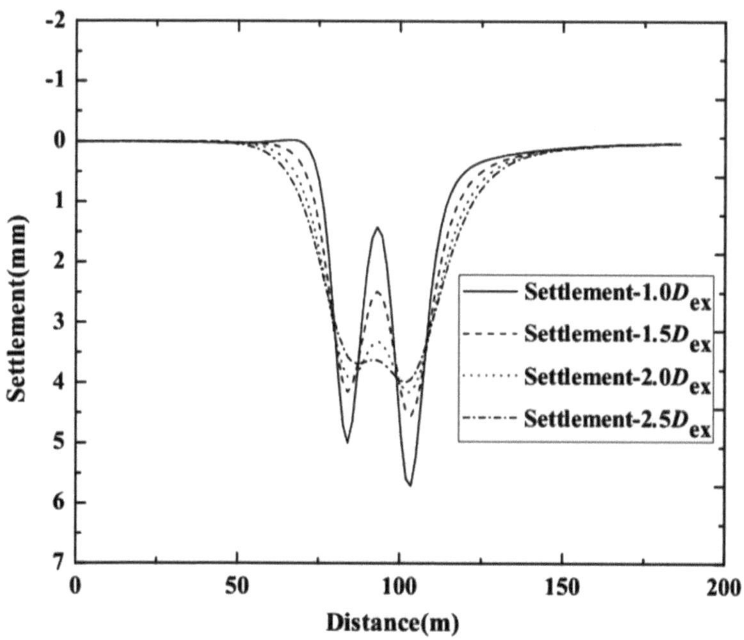

Fig. 6. Comparison of subsidence profile for twin tunnels (FLAC 2D) at various cover depths

Since the tunnels are constructed sequentially it will be prudent enough to understand whether construction of new tunnel, would cause any detrimental effect on the already constructed liner (in terms of forces). For this a pillar width sensitivity analysis is performed. This is performed by varying the pillar width between the twin tunnels from $0.5D_{ex}$ to $2D_{ex}$ at equal $0.5D_{ex}$ increments for different cover depths. The increase in thrust on liner obtained from the above sensitivity analysis is shown in Fig. 7. The main deduction from this figure is that the pillar width effect in terms of tunnel liner forces are more significant for thrust forces than for moments. Thus the effect of thrust forces on constructed liner due to adjacent tunnel construction has to be accounted although not significant (7.23% increment at $0.5D_{ex}$ pillar width). This increment of thrust forces on the constructed liner, especially at high cover depths, could be attributed to the load sharing phenomenon between twin tunnels. At this point it should be mentioned that for any tunnel (single), with an increase in cover depth, main structural force impacted could be normal thrust, followed by moments as shown in Fig. 8. This pillar width sensitivity analysis, especially focusing on impact of liner forces on the constructed liner by adjacent excavation will cause less impact on Delhi metro tunnels as they are generally constructed at a pillar width of $2D_{ex}$ or more.

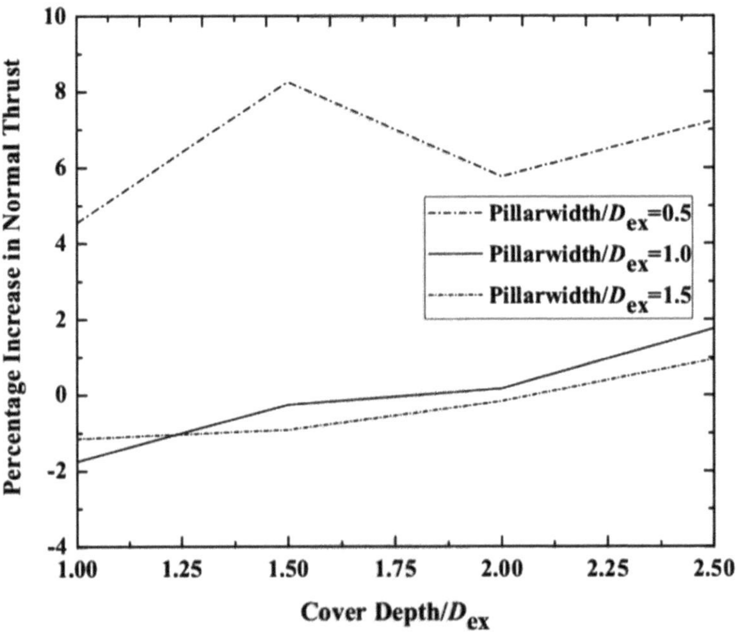

Fig. 7. Relationship between pillar width and normal thrust (constructed liner) at various cover depths

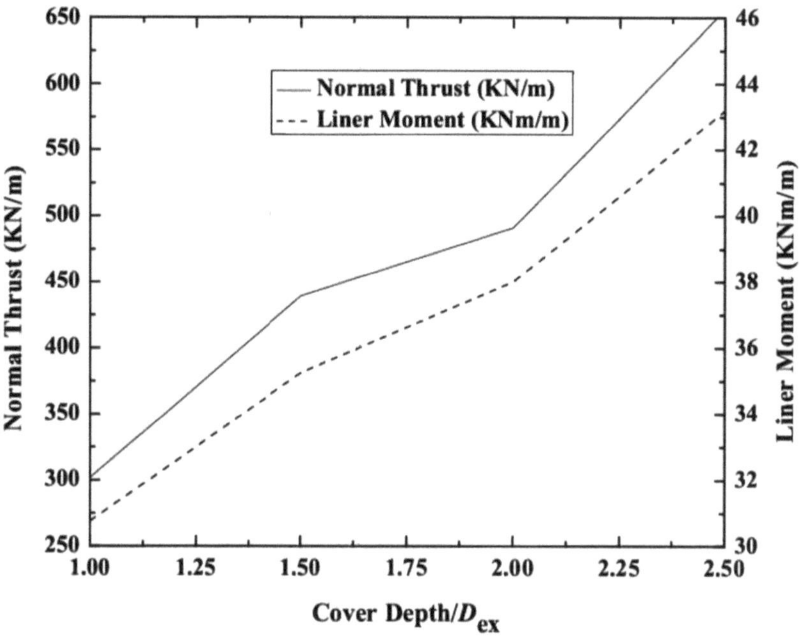

Fig. 8. Relationship between normal thrust and moment for a tunnel at various cover depths

Apart from pillar width sensitivity analysis focusing on constructed tunnel liner forces mentioned above, it will be appropriate to find out the variation of surface displacements with pillar width at various cover depths. It could be foreseen that with the decrease in pillar width, excavation volumes come closer resulting in a deeper settlement trough compared to the ones with higher pillar widths (Fig. 9).

Another significant issue during adjacent tunnel construction near a "constructed" tunnel could be pillar width stability or strength of the pillar between them. This effect could be appreciated by pillar width stability study focusing on "strength factor" between the tunnels. Alike the pillar width sensitivity analysis, this pillar width "stability "study is also performed for twin tunnels at different pillar widths for varying cover depths. Finite element analysis using *Roc science program Phase 2.0 v9*, is used for the computation of strength factor. It has been found that the critical stage occurs during the stress relaxation of second tunnel. It has been found that pillar width stability is an issue of concern at high cover depths (especially at low pillar widths, like $0.5D_{ex}$), where as a significant stability factor (around 2) could be obtained at higher pillar widths ($2D_{ex}$). The results are shown in Fig. 10. For this reason it will be appropriate to have a pillar width of '$2D_{ex}$' for metro tunnels. Now it will be appropriate to perform pseudo-static analysis of these structures to appreciate its effect in the post construction stage. As time history dynamic analysis requires elaborate computational resources, the practice is to conduct a pseudo-static analysis. This method as per *IS 1893*, is an inertia method of analysis where the contributing mass by the earthquake is incremented by some calculated amounts compared to the static case. Details are discussed below.

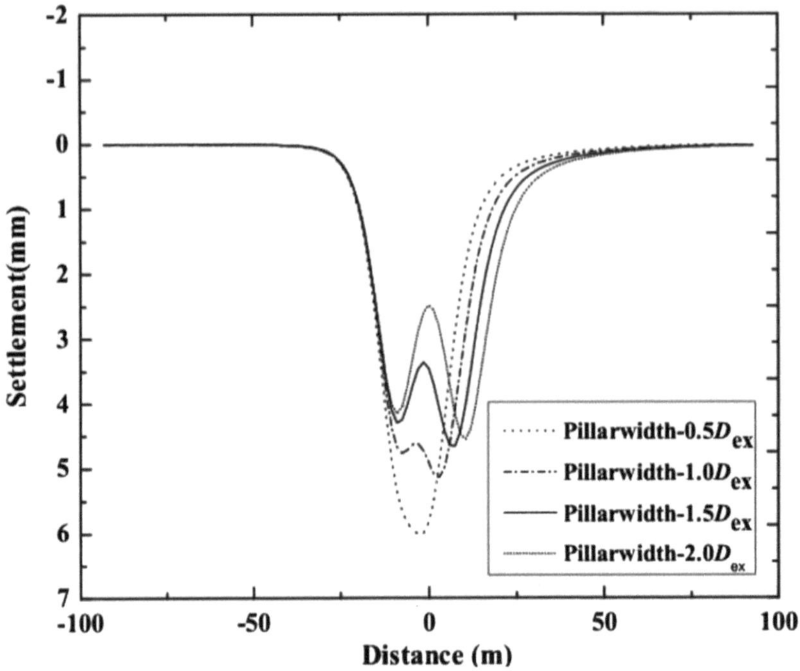

Fig. 9. Relationship between surface settlements and pillar width for a cover depth = $1.5D_{ex}$

1.3 Pseudo Static Analysis for Twin Tunnels

Pseudo static analysis is performed for twin tunnels as per IS 1893 (2002). As per codal provisions, for Delhi in seismic zone *IV,* a zonal factor (*Z*) of 0.24 g could be adopted. An importance factor, *I* of 1.5 (for important structures), a response reduction factor, *R* (ordinary reinforced concrete structures) of 3.0 and average response acceleration coefficient (S_a/g) of 2.5 (maximum spectral response recommended) is used for the calculation of horizontal seismic coefficient, A_h. This obtained value of A_h, is reduced for underground structures as per the recommendations in *IS 1893*. For structures at 30 m depth, A_h could be reduced by 50% of its original value at the surface. Hence for different cover depths appropriate values of A_h have been interpolated in this case. As per *IS 1893*, *Clause 6.4.5*, design acceleration spectrum for vertical motions may be taken as two thirds of design horizontal acceleration spectrum. Such vertical coefficients were considered both in the upward and downward direction, to deduce the governing case. For all the analysis considered, a pillar width of $2D_{ex}$ is considered as it has been the general case with most of the Delhi metro tunnels.

The main effect (in terms of percentage increase from static case) on maximum normal thrust and maximum moment are considered. As expected significant increase for these parameters are found at low cover depths as design horizontal seismic

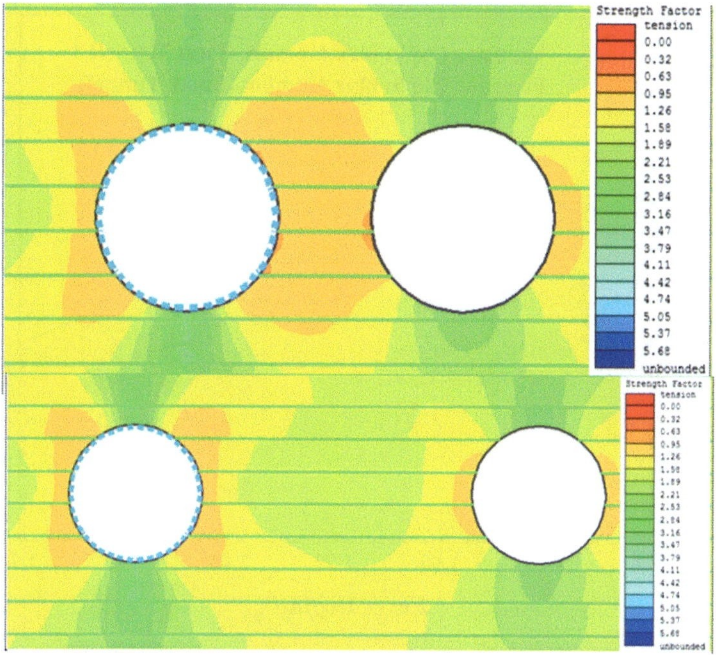

Fig. 10. Relationship between strength factor distribution and pillar width ($0.5D_{ex}$, $2D_{ex}$) for a cover depth = $2.5D_{ex}$

coefficient (A_h) increases towards the surface resulting in more inertial loads. The increase in moment and normal thrust due to pseudo-static force are given in Figs. 11 and 12 respectively. It is seen from these figures that at low cover depths (for instance $1D_{ex}$), normal thrusts were found to increase by as large as 48%, whereas maximum moments were found to increase by as large as 83% for the second tunnel. However, the percentage increase is relatively lesser for first tunnel. This indicates that even in pseudo-static analysis, the effect of twin-tunnel construction needs to be incorporated. It could be seen that in terms of thrust and moments latter tunnel gets affected more in terms of force increments. Apart from the increase in structural forces for tunnels, there will be an apparent phase shift (Fig. 13) for moments. Also surface displacements are computed for this study and are shown in Fig. 14. It is observed from Fig. 14 that the surface settlement profiles are distorted due to large pseudo-static forces in both horizontal and vertical direction. To understand the real response on surface displacements and liner forces it is recommended to perform a true dynamic analysis as a future study.

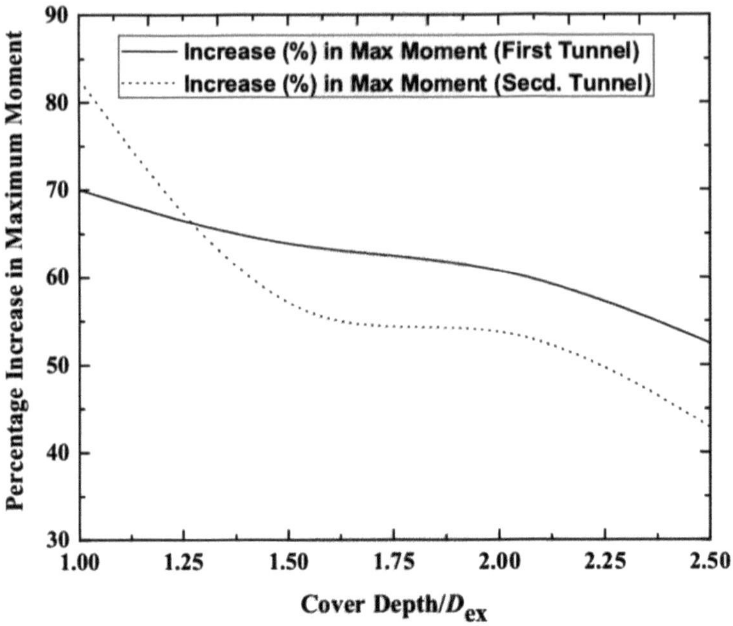

Fig. 11. Relationship between maximum moment and cover depth for both the tunnels in pseudo-static condition

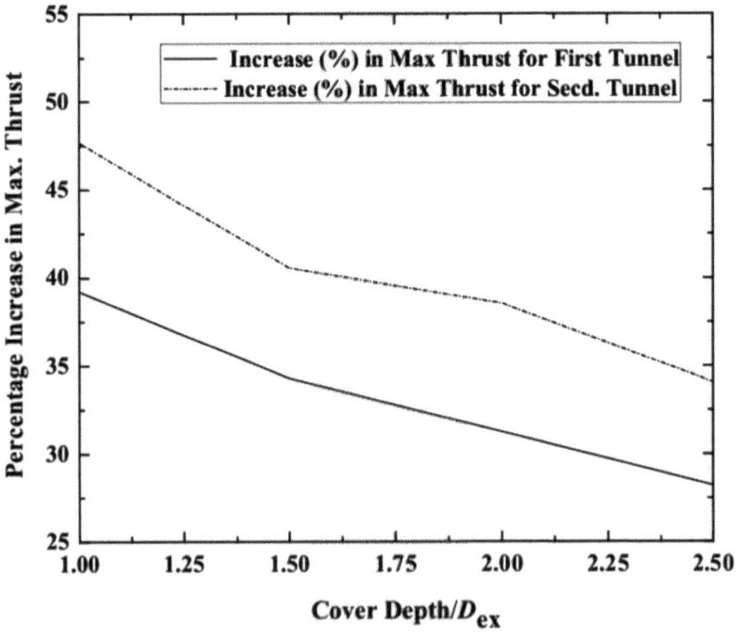

Fig. 12. Relationship between maximum thrust and cover depth for both the tunnels in pseudo-static condition

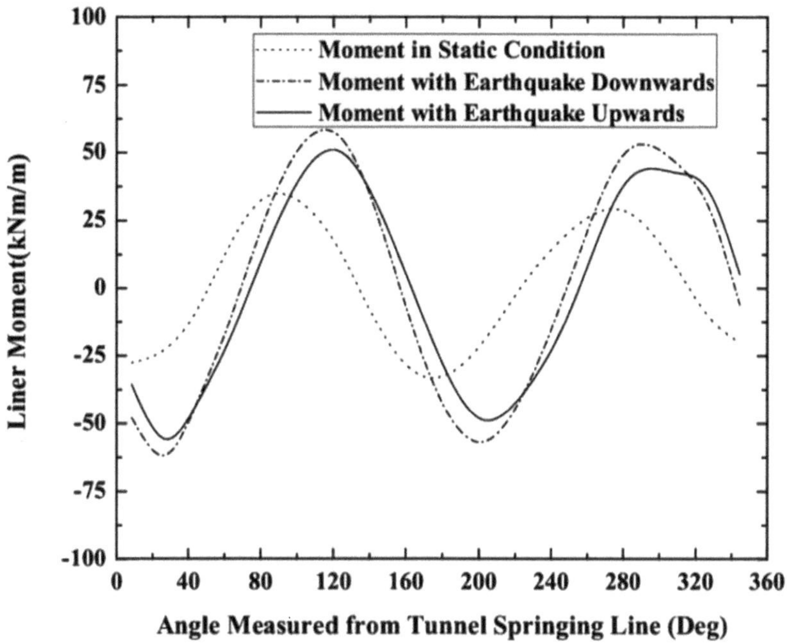

Fig. 13. Comparison between moment in static condition and pseudo-static condition.

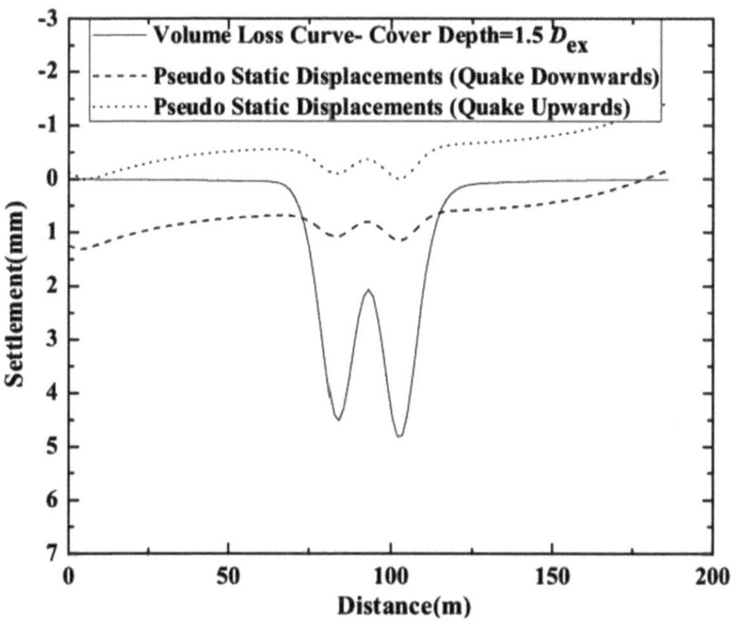

Fig. 14. Surface displacement profiles by pseudo-static analysis (cover depth = $1.5D_{ex}$)

2 Conclusions

Based on the above analysis, the following conclusions are arrived at:

1. Pillar width sensitivity and stability analyses, indicates that for Delhi metro tunnels it is safe to have minimum pillar width of $2D_{ex}$ or more, for all cover depths.
2. Surface settlements for twin tunnels calculated by superposition method are less reliable as twin tunnel settlement curves are generally asymmetric. So especially for twin tunnels numerical methods are appropriate for settlement deduction.
3. It is known that for a constant volume loss, settlement trough tends to flatten with increase in cover depths resulting in wider settlement troughs. This is true for both twin tunnel and single tunnel excavation problems.
4. It is observed that at low cover depths (for instance $1D_{ex}$), increase in thrusts and maximum bending moment are significant when the pseudo static forces are considered. Moreover, the percentage increase is also affected by twin-tunnel construction, which needs to be considered in pseudo-static analysis.

References

Anon: A Report on Seismic Hazard Microzonation of NCT Delhi on 1:10,000 scales, National Center for Seismology, Ministry of Earth Sciences, Government of India (2015)

Blom, C.B.M.: Concrete Linings for Shield Driven Tunnels. Technische Universiteit Delft, Delft (2009)

Bobet, A.: Analytical solutions for shallow tunnels in saturated ground. J. Eng. Mech. **127**(12), 1258–1266 (2001)

Chapman, D.N., Ahn, S.K., Hunt, D.V.: Investigating ground movements caused by the construction of multiple tunnels in soft ground using laboratory model tests. Can. Geotech. J. **6**, 631–643 (2007)

Gomes, R.C.: Seismic behaviour of tunnels under seismic load, Masters thesis. Instituto Superior Technico, Technical University of Lisbon (2000)

Hanumanthrao, C., Ramana, G.V.: Dynamic soil properties for microzonation of Delhi, India. J. Earth Syst. Sci. 719–730 (2008)

IS 1893: Indian Standard Code Practice - criteria for earthquake resistant design of structures, Part 1. Bureau of Indian Standards, New Delhi (2002)

Kim, S.H., Burd, H.J., Milligan, G.W.: Interaction between closely spaced tunnel in clay. In: Proceedings of the International Symposium on Geotechnical Aspects of Underground Construction in Soft Ground, pp. 543–548 (1996)

Lake, L.M., Rankin, W.J., Hawley, J.: Prediction and effects of ground movements caused by tunneling in soft ground in urban areas. CIRIA Project Report 30, Construction Industry Research and Information Association (1992)

Lysmer, J., Kuhlemeyer, R.L.: Finite dynamic model for infinite media. J. Eng. Mech. Div. 859–877 (1959). ASCE

Peck, R.B.: Deep excavations and tunneling in soft ground. In: Proceedings, Seventh International Conference on Soil Mechanics and Foundation Engineering, Mexico City, pp. 225–290 (1969)

Seed, H.B., Idriss, I.M.: Soil moduli and damping factors for dynamic response analysis. Technical report, EERRCC-70-10, University of California Berkeley (1970)

Abu Hamour Tunnel Phase I the First TBM Tunnel in Qatar: The Art of Tunneling in a New World

J.B. Stypulkowski(✉) and F.G. Bernardeau

CDM Smith, Doha, Qatar
{stypulkowskijb, BernardeauFG}@cdmsmith.com

Abstract. Abu Hamour (Musaimeer) Surface & Ground Water Drainage Tunnel – Phase I (AHSO) is a 9.5 km long, 3.7 m ID storm water tunnel. The writers created comprehensive geotechnical data collection system on the tunneling project in weak/soft rocks which has been successfully implemented. Results of geological mapping have been presented to show how they correlate to TBM performance. The authors make an attempt to correlate common classification systems: Q_w and RMR for rock mass commonly found in Doha, Qatar.

1 Introduction

Abu Hamour (Musaimeer) Surface & Ground Water Drainage Tunnel – Phase I (AHSO) is 9.5 km long with 3.7 m ID storm water tunnel about 30 m below ground surface. Completion of the project is scheduled for 2017. The Owner (or Employer), PWA-ASHGHAL, is the designated Engineer, who is represented by the Engineer's Representative CDM Smith. CDM Smith design team provided engineering design review, verification and approval services for all the design produced by the Contractor's team. CDM Smith is also the Construction Manager on this project.

1.1 Project Overview

Integral to Phase I of the project, are 19 access shafts (AS) located at about 500 m on center. Some of these shafts are drop shafts that facilitate runoff inflows along the route of the tunnel. Most of the shafts are offline therefore requiring the construction of 15 adits, and application of sequential excavation tunneling methods. The tunnel runs from an existing access shaft AS23 to a retrieving shaft located within footprint of a future pumping station at the coastline to be constructed under Phase II of the project. There are 3 online shafts, 1 existing shaft, 15 offset shafts (which are connected to the main tunnel through adits), and 6 offline branch shafts (which are terminal shafts facilitating connections accomplished by microtunneling).

Access Shaft AS11 a launching shaft, centralized along the alignment, was used to launch two Tunnel Boring Machines (TBMs) mining in opposite directions. Access Shaft AS11 provided access for tunnel construction; it was 32 m in length, 10 m wide and 31 m deep. After completion of construction a smaller 5.4 m diameter permanent

S. Agaiby and P. Grasso (eds.), *Engineering Challenges for Sustainable Underground Use*, Sustainable Civil Infrastructures, DOI 10.1007/978-3-319-61636-0_11

shaft will be constructed within the temporary shaft and the excess space around it backfilled. Two shafts of 10 m diameter were intended as TBM reception shafts at each end of the project: AS22R and PS-TBM recovery shaft east, located inside future pump station footprint. For other publications related to the project please refer to Stypulkowski et al. 2013, 2014a, b, 2016; Siyam et al. 2014, 2015.

2 Geology

2.1 Regional Geology

The project area is geologically a part of the Arabian Gulf Basin. It forms a part of the Arabian shelf between the Arabian shield and Iranian mobile belt. Thickness of sediments in the Qatar region is estimated to be about 10 km (Abu Zeid 1991; LeBlanc 2008). The post Cretaceous sedimentation is basically a sequence of shallow marine limestone with occasional shale and evaporates in a shallow basin. The formations encountered in Doha region comprise of Quaternary marine, aeolian and sabkha deposits. They are: Rus of Lower Eocene, Lower Dammam and Upper Dammam of Middle Eocene and Lower Dam of Lower Miocene. We encountered following rock domains listed from ground surface down (see Fig. 1): Simsima Limestone (represented by 3 shades of blue on Fig. 1) – which contains Dolomite of Upper Dammam formation, Midra Shale (yellow) of Lower Dammam formation and Rus (grey) of Lower Eocene is below Midra. Water making horizons are marked in green.

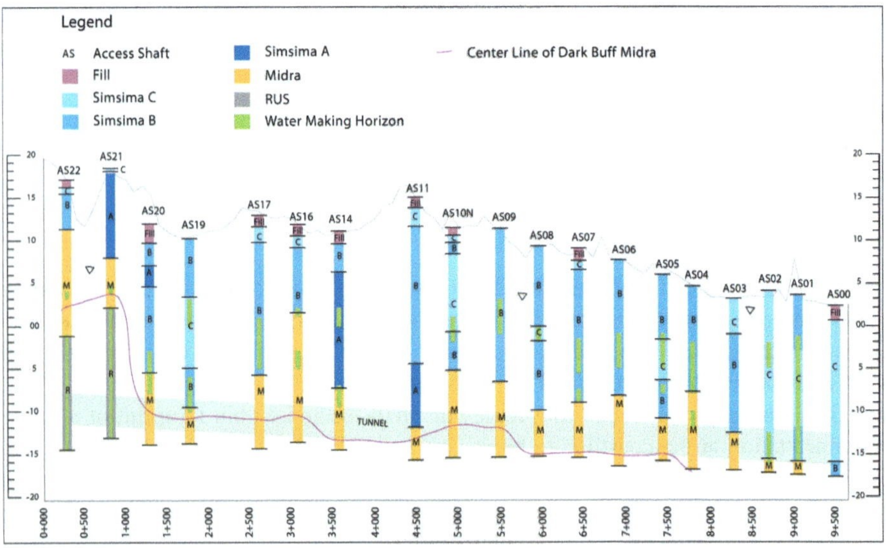

Fig. 1. Interpreted geology from shaft mapping (both scales in m)

2.2 Geology Encountered

Integral part of supervision of the field activities is observation of rock as is exposed during excavation. Use of the shielded TBMs with segmental lining prevented continuous geological mapping. Rock mass therefore was properly surveyed only at all shaft locations spaced about 500 m apart. The objective of the wall mapping program was to establish the large scale rock mass characteristics by geotechnical observations (Q, RMR) of the exposed rock face. Mapping procedures and results have been described in detail in Pathak et al. 2015.

2.3 Intact Rock Properties

Laboratory tests were conducted on rock samples obtained from borings before construction. Uniaxial compressive strength (UCS) test results range between 2–65 MPa (median 15 MPa). Tensile strength from Brazilian tests range between 0.1–9.1 MPa (median 1.7 MPa). Is (50) from point load testing ranges between 0.03–7.5 MPa (median 1.3 MPa). Young's modulus from UCS testing range between 0.4 and 49 GPa (median 4.2 GPa), from pressuremeter testing in initial loading range 79–5,310 MPa (median 389 MPa) while down-hole seismic test results range is 1–4.3 GPa (median 2.4 GPa).

2.4 Rock Mass Classification

The writers decided to look at two systems at the same time. The Q-method (NGI 2013) was the first choice since it allows for calculation of wall specific Qw directly applicable to shaft walls. Since only a few cases from weak and weathered rock were used in its development, other methods had to be considered. The RMR system (Bieniawski 1989) while it was developed for tunnels it is commonly used by designers since it is related to GSI which in turn is often used as an input for the Hoek-Brown failure criterion. The mapping results indicate that rock quality in terms of Qw range from extremely poor to poor. The calculated correlation between Q and RMR results are shown on Fig. 2 and are similar to ones presented in literature (Hoek's corner). Mapping process, results and analysis have been described in detail in Pathak et al. 2015 and are summarized in Table 1. Predictably original assessment when compared with shaft mapping results has been generally conservative. Mapping results when compared with relogging of the available shaft borings was generally optimistic. When mapping results were compared with re-logging effort in 72%-RMR, 61%-Q, results were similar, reflecting the logging bias. The logger had seen rockmass intact in the shafts so he knew what to look for during re-logging. In 18%-RMR, 22%-Q mapping turned up more optimistic than relogging and only about 10%-RMR, 18%-Q largely for near the surface results, mapping showed worst rock that otherwise would have been predicted.

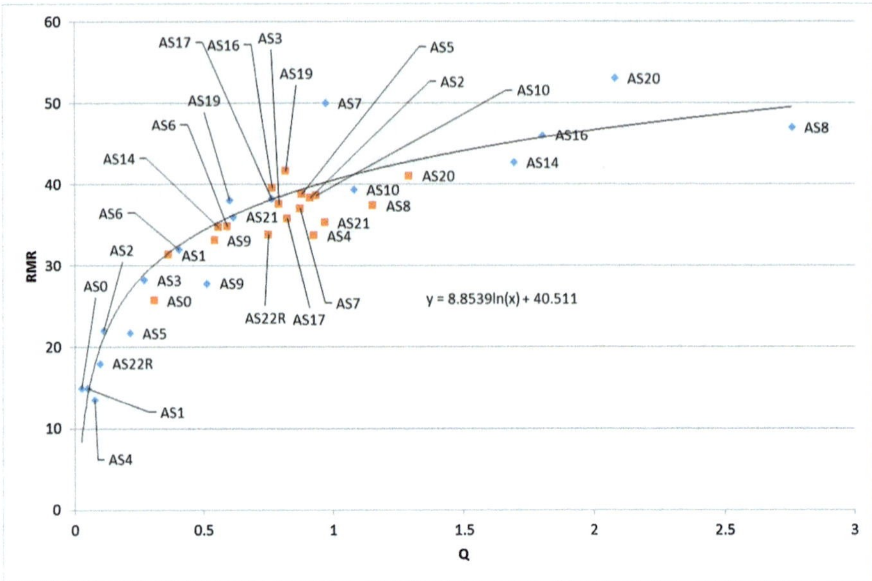

Fig. 2. Q and RMR correlation from relogging (orange) and shaft wall mapping (blue) at the tunneling/adit horizon (eastern & western drives combined)

Table 1. Statistical summary for input parameters and classification results

Classification system parameters/results	Original assessment - Median	Re-logging - Median	Shaft mapping max - Median	Shaft mapping min - Median
RQD	55	37.3	50	40
Joint number	6	20	15	20
Joint roughness	3	4	4	3
Joint alteration	3	4	3	2
Joint water	0.66	0.66	0.66	0.5
Q	1.232	0.58	2	0.792
Rating for UCS	2	2	2	1
Rating for RQD	13	8	8	3
Rating for spacing of discontinuity	5	8	8	5
Rating for condition of discontinuity	9	10	10	0
Ground water rating	10	7	10	7
RMR	34	37	38	18

2.5 Muck

Limited laboratory testing of the muck has been conducted. Test results shown on Fig. 3 indicate clay content between 23% and 35% and silt content varies from 9 to 16% and PI from 57 to 36 and LL 10% to 130%. However, gradation curve showing passing the 0.075 mm sieve is 44 and 39%. For eastern sample with 56% sand and/or gravel, 44% silty clay limits plot as silt of high plasticity, elastic silt (MH) and classifies as clayey sand (SC) or clayey gravel (GC) in unified system. For western sample with 61% sand and/or gravel, 39% silty clay limits plot as MH and classifies as SC or GC in unified system as well.

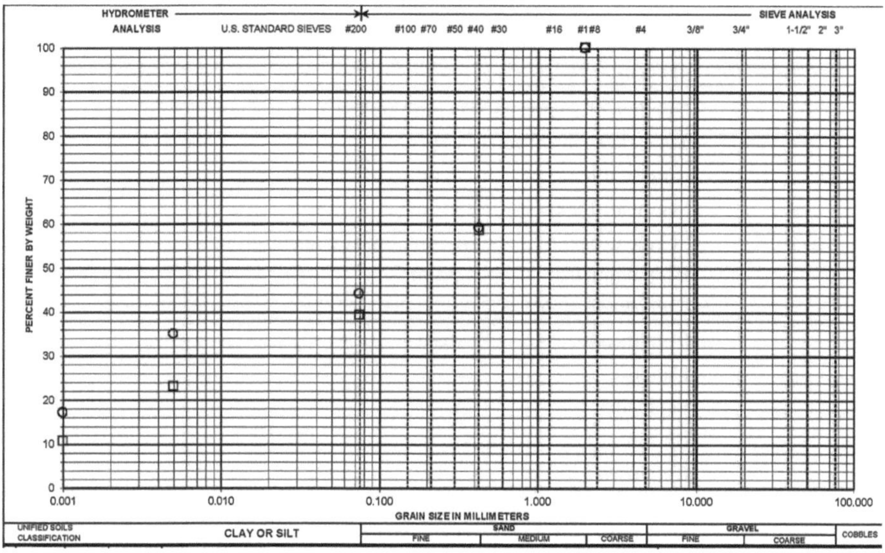

Fig. 3. Sieve & Hydrometer test results for muck samples

3 Examination of TBM Data

The excavation was carried out by two TBMs with rotating cutter-head which was fitted with cutting wheels and cutting tools (buckets and slab cutters). The excavated material was collected by the "buckets" and transported through the openings provided between the cutters into the excavation chamber. A screw conveyor extracted the spoil from the chamber and discharged it onto the belt conveyor installed on the TBM back which in turn offloaded the material into muck skips. The muck skips (total 26) were operated by a diesel powered locomotives (total 7) which traveled on the tunnel rail track. The permanent lining of the tunnel consists of dowelled, pre-cast concrete segments reinforced with steel fibers. See Table 2 for additional details. Monitoring of the TBM excavation processes was provided in real-time by Herrenknecht (HK) made

Operation (HK) and Guidance (VMT) Systems. The drive data was automatically collected by the TBM Data Acquisition and Evaluation (DAE) System in CSV format. The HK protocol was used to print predefined reports. The contractor kept their own paper records as well.

Table 2. TBM specifications

Design parameters	
Curve radius	300 m
Gradient	0.05%
UCS	2-65 MPa
Hydrostatic pressure	2* bars
Segmental lining	
Number of segments	Per ring 6+key
Total rings installed	7198
Ring width	1.3 m
Segment thickness	250 mm
Segment backfill	Bi-component cement grout
Bore diameter	4.52 m
Cutterhead style	
Cutterhead	Mixed
Cutters	17
Scrapers	75
Buckets	8
Cutterhead drive	
Cutterhead speed	0-4, 5 min^{-1}
Nominal torque	2167 kNm
Maximum thrust	20891 kN
TBM conveyor	
Screw conveyor dia.	600 mm
Weights and dimensions	
Total length	124.5 m
TBM weight	330 ton
TBM core weight	172 ton

* TBM design pressure was 3 bar.

There were 62 interventions at the face for both drives, 45 involved disc and scraper replacement including four where rippers replaced discs. Overall disc use per meter of tunnel varied between 0.0069 and 0.23 (mode 0.007) and histogram of frequency distribution for both drives is shown on Fig. 4.

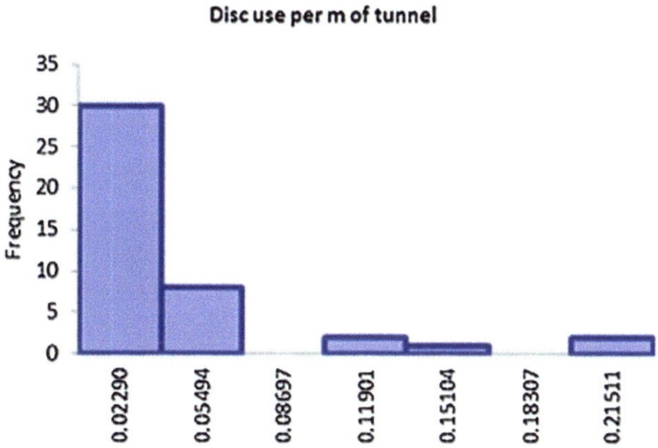

Fig. 4. Frequency distribution of the disk use per meter of the tunnel

3.1 TBM Penetration Considerations

It is expected that factors impacting the Penetration Rate (PR) in soft ground will be different than in hard rock for EPB TBM. The PR in hard rock is linked to rock hardness etc. while soil conditioning determines PR in soft ground. The TBM performance prediction models for hard rock has been described in detail in Hassanpour 2011. Maher in his on-line article suggested that there is a relationship between PR and surfactants used in soft ground. In this paper authors examine applicability of both approaches in weak/soft rocks of Doha. Significant difference in mode terms (see Table 3) for TBM cutterhead torque and thrust force per cutter was noted between the drives and it was 36% and 28% lower for the east drive respectively. Therefore, resulting penetration rate for eastern drive was 40% faster.

Table 3. Statistical summary of TBM performance parameters

	Western drive			Eastern drive		
	Mean	Mode	Std. dev.	Mean	Mode	Std. dev.
TBM cutterhead torque (kNm)	1,226	1,408	261	1,149	901	243
TBM thrust force per cutter (kN)	382	407	62	341	292	68
TBM penetration rate (m/hr)	2.39	2.18	0.74	2.86	3.07	0.77

The authors looked at correlations between the torque (kNm) and thrust per cutter (kN) vs. penetration (mm/rev) for long sections as well as areas around the shafts. In general, no increase of penetration can be associated with thrust and torque increase typically observed in hard rock. This kind of behavior has been observed in highly

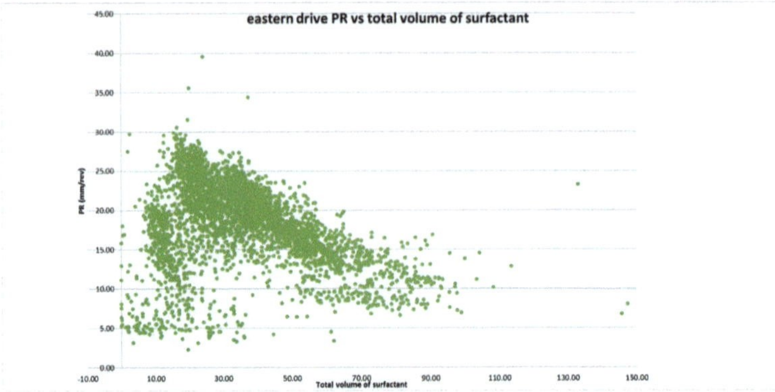

Fig. 5. Relationship between total volume of surfactants (BASF - MasterRoc SLF33) vs. penetration rate (PR) in (mm/rev) for eastern drive

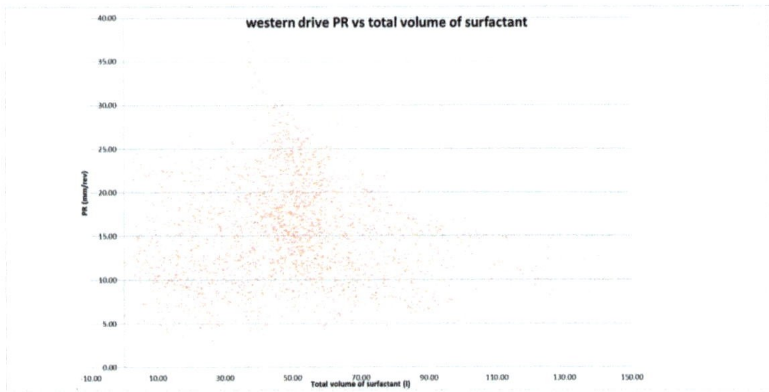

Fig. 6. Relationship between total volume of surfactants (MAPEI - Polyfoamer FP/LL) vs. penetration rate (PR) in (mm/rev) for western drive

fractured rock by Avunduk et al. 2012. However, the penetration rates of both TBMs show clear relationship between increases in surfactant volume in response to decreasing penetration rates on Figs. 5 and 6.

In order to provide desired consistency of the muck for most of the drive additional ground water had to be pumped in together with surfactants diluted by service water. As shown on the Fig. 7 significant quantities of water were required for all the drives in an aquifer below layer of Dark Buff Midra. While volume of the surfactant shows predicable relationship with the PR ground water volumes don't. Lack of relationship for ground water can be attributed to its manual set up at TBM control while surfactants were set in semi-automatic mode.

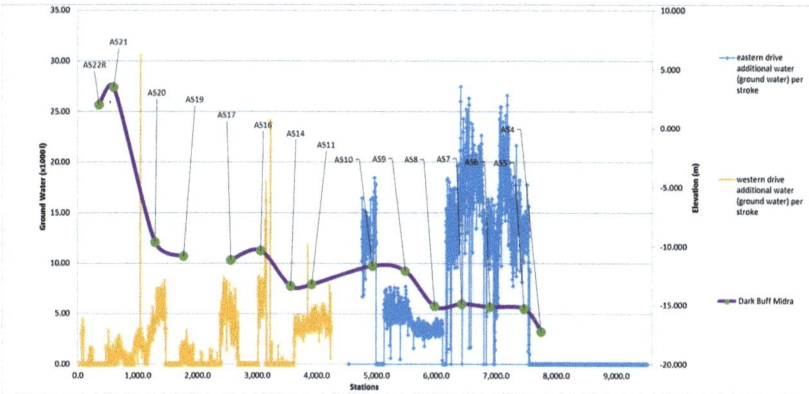

Fig. 7. Ground water pumped in directly to TBM for both drives

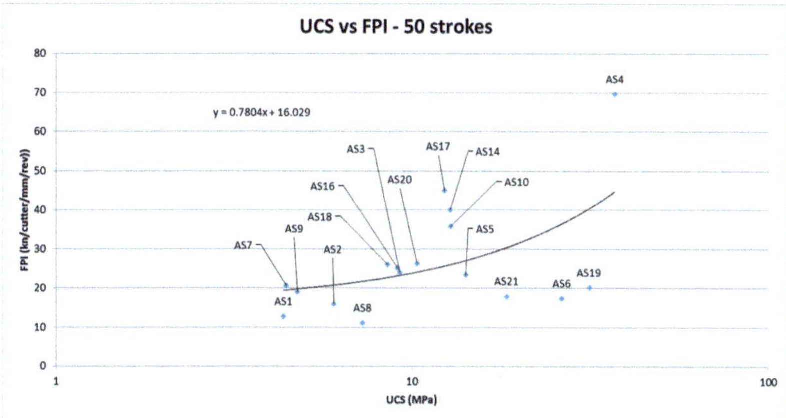

Fig. 8. Results of UCS and FPI correlations for 50 strokes around the shaft locations only

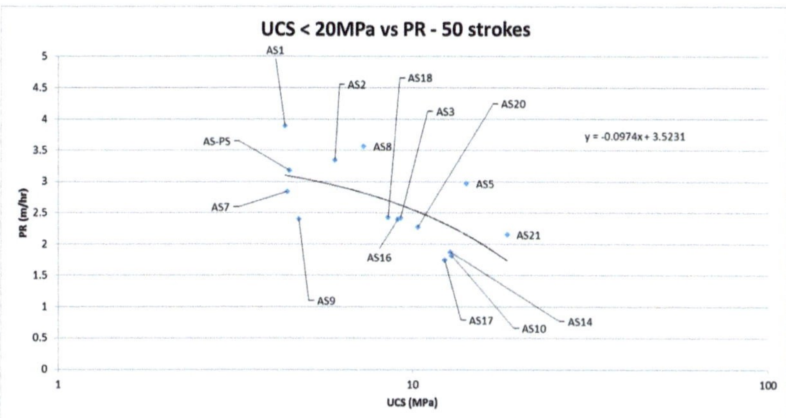

Fig. 9. UCS vs Penetrations in Shaft Areas for UCS < 20 MPa

3.2 UCS and Penetration in Weak/Soft Rocks

Sample preparation for weak rock materials is more difficult compared to that of hard rock materials. Often methods of sample preparation and laboratory testing for weak rocks are similar to soils. It has been reported (Stypulkowski et al. 2014) that it is often hard to obtain a sufficient number of samples with an L/D ratio bigger than 2, as the drilling program into weak rock is usually difficult. This also leads to collection of best samples for testing not the representative ones. Therefore, the relationships between UCS and penetration, torque and thrust are shown for access shafts (AS) on Figs. 8, 9, 10 and 11 for samples less than 20 MPa. Only for adjusted data-base consisting of test results from tunneling horizon, correlations with UCS has been successfully established showing expected trends.

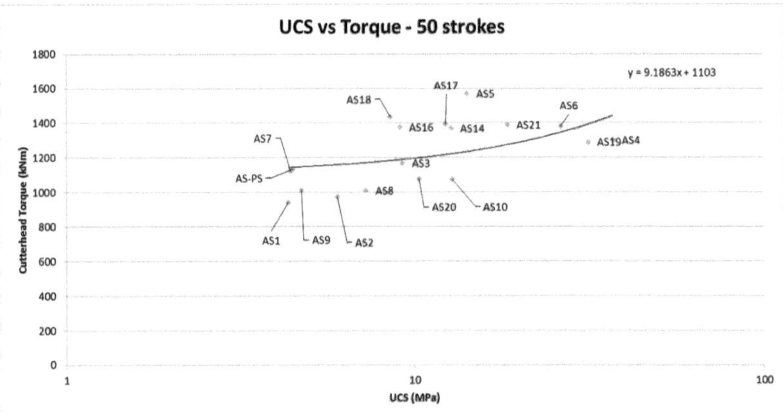

Fig. 10. UCS vs. Torque in shaft areas

3.3 Penetration Predictions

The latest trend in TBM performance prediction using empirical models often relay on historical field data in various ground conditions. The authors used published correlations in Hassanpour et al. 2011 to predict FPI and compare it with average of 50 strokes obtained from around the shafts. The results are presented in Fig. 12 and correlation for selected parameters (UCS, RQD, RMR & Q) appear to be poor. Actual average FPI rates were between 40% and 65% lower from actual. It can be therefore concluded that hard rock approach in FPI prediction can't be used in weak/soft rocks of Doha without adjustment.

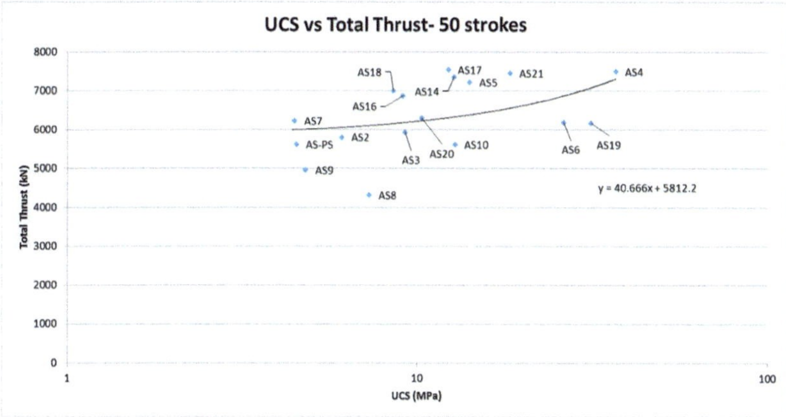

Fig. 11. UCS vs. Total thrust in shaft areas

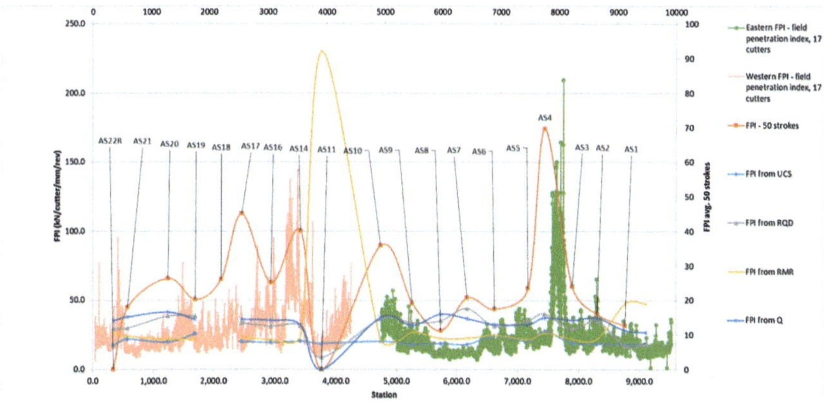

Fig. 12. Field Penetration Index (FPI – (kN/cutter/mm/rev) for both drives

4 Conclusions

The writers created and implemented comprehensive geotechnical data collection system on the tunneling project in weak/soft rocks which has been compared with data collected by the TBMs. Since geotechnical investigations were conducted largely around the frequently located shafts along the alignment the TBM data was delineated to areas around the shafts. At this time, it has been concluded that FPI predictions using hard rock models for rockmass commonly found in Doha would underestimate actual penetration rates. It appears that weak/soft rock when analyzed for TBM performance should be treated as soft ground.

Acknowledgments. The writers acknowledge Salini-Impregilo for successfully completing two TBM drives.

References

Abu Zeid, M.M.: Lithostratigraphy and framework of sedimentation of the subsurface Paleogene succession in northern Qatar. Arabian Gulf. N. Jb. Geol. Paliaiont. Mh., 191–204 (1991)

Avunduk, E., Copur, H., Bilgin, N., Balci, C., Tumac, D.: Effect of some geotechnical properties on TBM performance. In: Eurock 2012 – The 2012 ISRM International Symposium, Stockholm, Sweden, 28–30 May 2012 (2012)

Bieniawski, Z.T.: Engineering Rock Mass Classification, 272 p. Wiley, New York (1989)

Handbook- Using the Q-System Rock mass classification and support design. NGI (2013). www.ngi.no

Hassanpour, J., Rostami, J., Zaho, J.: A new hard rock TBM performance prediction model for project planning. Tunn. Undergr. Sp. Technol. **26**, 595–603 (2011)

LeBlanc, J.: A Fossil Hunting Guide to the Tertiary Formations of Qatar, 1st edn. Middle East (2008)

Maher, J.: A Machine Learning Approach to Predicting and Maximizing Penetration Rates in EPB TBMs (online version undated)

Pathak, A.K., Stypulkowski, J.B, Bernardeau, F.G.: Super-vision of engineering geological activities during construction of Abu Hamour surface and ground water drainage tunnel Phase-1 Doha, Qatar. In: International Conference on Engineering Geology in New Millennium at IIT, New Delhi, 27–29 October 2015, pp. 467–486. Special Publication, J of EG, October 2015

Siyam, A.F.M., Stypulkowski, J.B., Bernardeau, F.G.: Improvement to longevity of tunnels in aggressive ground conditions in the Middle East. In: Arabian Tunnelling Conference and Exhibition, Abu Dhabi, UAE, 9–10 December 2014, pp. 163–185 (2014)

Siyam, A.F.M., Bernardeau, F.G.: Structural and construction challenges for shafts on Abu Hamour surface and groundwater drainage tunnel. In: Arabian Tunnelling Conference and Exhibition, Dubai, UAE, 23–25 November 2015 (2015)

Stypulkowski, J.B., Siyam, A.A.F.M., Bernardeau, F.G., Al Kuwari, N.G.: Abu Hamour drainage tunnel, first TBM mined tunnel in Doha, Qatar. In: The First Arabian Tunnelling Conference & Exhibition, Dubai, pp. 300–314 (2013)

Stypulkowski, J.B., Bernardeau, F.G., Sandell, T.D.: Mechanized tunneling technologies for weak rocks of Middle East/Persian Gulf. In: ATC 2014 Arabian Tunnelling Conference and Exhibition, Abu Dhabi, UAE, 9–10 December 2014, pp. 187–199 (2014a)

Stypulkowski, J.B., Pathak, A.K., Bernardeau, F.G.: Abu Hamour, TBM launch shaft – a rock mass classification attempt for a deep shaft in Doha, Qatar. In: EU-ROCK 2014, ISRM International Symposium, Vigo, Spain, 27–29 May 2014, p. 117. CRC Press, Taylor & Francis Group (2014b)

Stypulkowski, J.B., Pathak, A.K., Bernardeau, F.G.: Engineering geology for weak rocks of Abu Hamour surface and ground water drainage tunnel Phase-1 Doha, Qatar. In: RARE 2016, International Conference on Recent Advances in Rock Engineering, Specialized Conference of ISRM in Bengaluru, India, 16–18 November 2016, pp. 85–90 (2016)

@Risk, Palisade Corporation, 31 Decker Road, Newfield, NY 14887

Analysis of Construction Scheme for Neighborhood Metro Tunnels with Two Different Inner Sections

Xiangxing Kong(✉)

The First Highway Survey and Design Institute of China Communications Construction Company Ltd., Xi'an, China
103121153@qq.com

Abstract. 2D-FEM was applied to simulate and analyze several construction schemes, of which shield method was adopted in left tunnel with small inner section and double-heading construction method or cross diaphragm method for right tunnel with large inner section. The results show that compared with the construction scheme that the tunnel with small inner section is firstly excavated, the one that tunnel with large inner section is first accomplished can evidently reduce the disturbance effect on surrounding rock, decrease the settlements of ground. Compared with CRD method, Double-heading construction method has advantages in utilizing the self-supporting ability of surrounding rock and supporting ability of preliminary lining, and also improving safety margin of the second lining.

1 Introduction

As the city subway, underground tunnels are taken to two equal sections (Bo et al. 2006), only meeting special engineering needs it is designed with two unequal sections, for example crossover, contact lines and stop (deposit) car line and so on. The tunnel with parallel non large section has carried out preliminary research. Through the establishment of numerical model (Jian et al. 2004), the influence of supporting mechanical behavior and stability of surrounding rock of unequal span highway tunnel are analysed; also, construction mechanics and surface settlement of double line tunnel with antisymmetric sections are also studied (Xiaorui 2009), but they are still need to further study.

In order to meet the two lane and the right line stop line section expanding engineering function, the left line of small section tunnel with diameter 6 m is constructed by shield method, large section tunnel right line height with 9.31 m width 11.24 m is constructed by the new Austrian tunneling method. A hole for the shield and another hole for the new construction scheme of NATM is selected.

For solving the general linear programming problem, the research and application of the small spacing tunnel begins. For example Shenzhen rail transit technology park to baidanzhou interval tunnel is about 0.5–1.2D (D as the tunnel span, similarly hereinafter) (Jianwu 2004). However, the engineering distance parallel to the tunnel is 4.247 m, only 0.38D, far less than the specified 1.0D current code for design of Metro in China, this will also bring engineering considerable difficulties and risks.

S. Agaiby and P. Grasso (eds.), *Engineering Challenges for Sustainable Underground Use*,
Sustainable Civil Infrastructures, DOI 10.1007/978-3-319-61636-0_12

2 Geology Conditions

There are artificial fill, quaternary late Pleistocene Aeolan loess soil, a new residual, late Pleistocene and Middle Pleistocene alluvial silty clay and sand soil, the interval tunnel passes through the upper layer, the new loess and paleosol in the upper, clay soil and sand soil in the lower, physical and mechanical properties of concrete around the base soil.

3 Numerical Simulation Model

The calculation program uses two-dimensional plane strain model. It is assumed that the soil with Drucker-Prager strength criterion and isotropic hardening elastoplastic constitutive model of soil (Xiangxing et al. 2011), the quadrilateral element mesh division, and the shield segment is simulated by beam element, advance grouting pipe reinforcement effect to improve the stability of the surrounding rock of the quadrilateral element simulation, the initial support and steel the grid element simulation, two lining is simulated by beam element, considering the initial lining and lining between the two waterproof board using contact element simulation. In order to ensure the accuracy of the dense units around the tunnel are used in the construction scheme of finite element model as shown in Fig. 1.

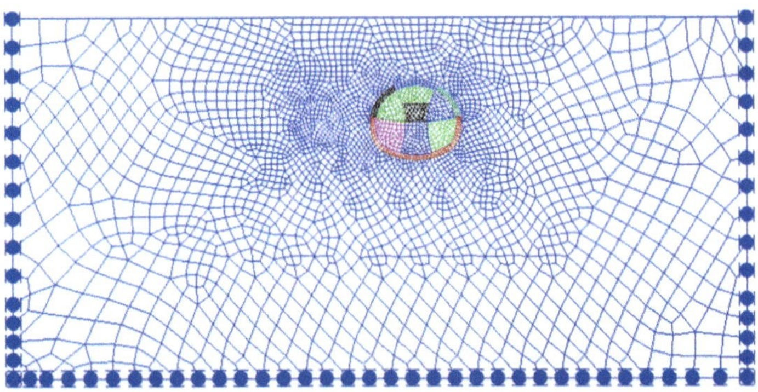

Fig. 1. Mesh of finite element model

According to the Saint Venant's principle and the actual needs of, calculation model for the horizontal direction is 4 times diameter, 40 m, vertical direction is 3 times diameter, 30 m. Vertical direction down to get to the surface. Model boundary conditions on the left and right sides of a given X displacement constraint, the bottom surface Y direction displacement constraint. According to the data of engineering geology, soil and calculation of related parameters are shown in Table 1.

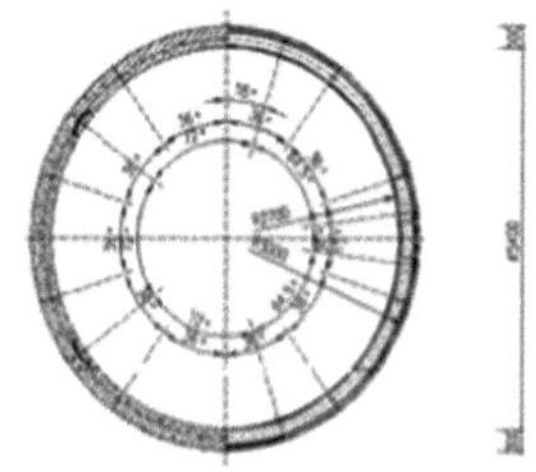

Fig. 2. Segment cross-section of shield method tunnel

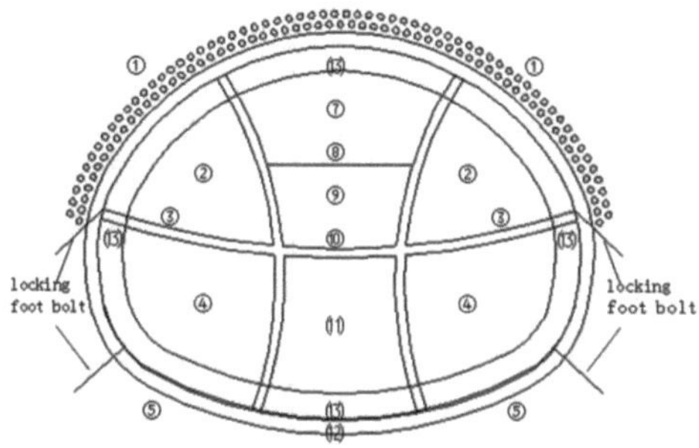

Fig. 3. Construction sequence of double-heading method

Table 1. Parameters of support structures

Soil/structure	Density/ ($kN{\cdot}m^{-3}$)	Elastic modulus/Mpa	Section area/m^2	Cohesion/KPa	Inner grazing angle/(°)
Soil	18.6	10.3		36	24
Shield lining	25	34500, $\eta = 0.8$	0.3	–	–
Advanced grouting soil	23	22.5	–	54	37.5
Primary lining	25	23000	0.35	–	–
Geogrid shotcrete	25	34500	–	–	–
Secondary lining	25	30000	0.5	–	–

4 Construction Schemes Simulation

The left line of small section tunnel excavation diameter is 6.0 m, C50 concrete segment thickness is 0.3 m, the impermeability grade of S10, specific ring lining section shown in Fig. 4. Large sections of the right line of NATM tunnel excavation is high 9.31 m, width 11.24 m, C25 lining thickness 0.35 m, HPB335 thick 0.25 m steel grille. In addition, in the loose loess shallow subway tunnel excavation, construction of pre grouting pipe in the tunnel arch can improve the condition of surrounding rock to

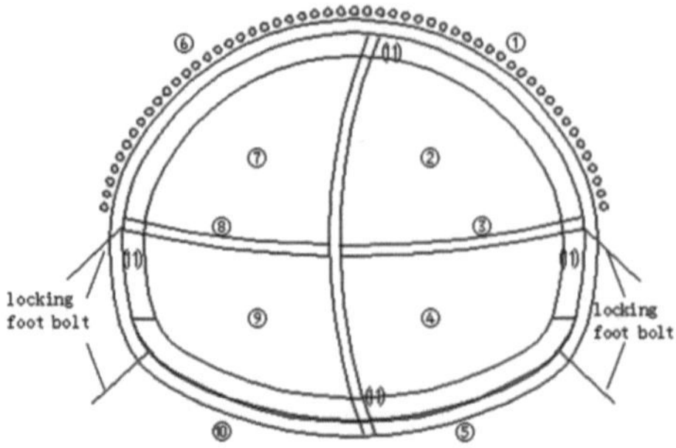

Fig. 4. Construction sequence of CRD method

ensure the stability of the tunnel face, and it is very important for increasing the stability of tunnel. The specific scheme is diameter φ 42 × 3.5 (thickness 3.5 mm) grouting pipe, length 2.5 m, angle 15°, circumferential spacing 0.3 m, vertical spacing of 1.0 m; pressure injection of cement - water glass grout, grouting pressure control in 0.6–3.5 MPa (Fig. 2).

Because the CRD method and the double side drift method is suitable for the large section and small spacing shallow buried loess tunnel. The total of 3 schemes of numerical simulation comparison with analysis to determine the interval tunnel construction scheme optimization which is the small section tunnel large section of shield method for double side heading method or CRD method in the construction method of construction order double successively through and large section. The following specific construction process dynamic simulation:

(1) Small section tunnel in left line firstly, the right line of large section tunnel by double side heading method finally.

At first, completed the small section tunnel left line, simulation step of shield construction into 2 steps: ① excavation; ② the application of segment lining. Then the big cross section tunnel in right line, double side heading method construction process simulation into 13 steps: the left and right up the steps: ① in the application of heading arch part of advanced small pipe grouting; ② excavation; ③ the vertical steel grille, shotcrete; ④ both sides heading down the steps followed in the excavation of the frame; ⑤ steel grille, shotcrete; ⑥ central heading up the steps in the application of arch part advanced small pipe grouting; ⑦ the excavation; ⑧ the vertical steel grille, shotcrete; ⑨ central drift step sequentially during excavation, removal step temporary bracing; ⑩ the vertical steel grille, shotcrete; central drift next step is ⑪ excavation; ⑫ sprayed concrete; ⑬ applied two lining (Fig. 3).

(2) Large section of double side heading method right tunnel firstly, shield small section of left tunnel finally.
(3) The right line of large section tunnel by CRD method firstly, small section tunnel in left line finally.

At first, the big cross section tunnel right line, simulation step construction with CRD method into 11 steps: Step 1 right sequentially as arch advanced small pipe grouting;

② excavation; ③ the vertical steel grille, shotcrete; ④ right down the steps in the excavation; ⑤ the vertical steel grille, shotcrete; ⑥ left on the application of step sequentially arch part of advanced small pipe grouting; ⑦ the excavation; ⑧ the vertical steel grille, shotcrete; ⑨ left down the steps in order to excavation; ⑩ the shotcrete lining ⑪ applied two times. Then completed the small section tunnel left line, simulation step of shield construction into 2 steps: ① excavation; ② the application of segment lining.

Schemes (1) and (2) are only difference on applied sequence of big cross section tunnel by double side heading method small section of tunnel by shield.

5 Results and Analysis

(1) Mechanics analysis for large section tunnel in the right line.

According to Table 2, scheme 2 produced in construction of initial support axis force is more than scheme 3, but the difference is small, more than scheme 1. As the hole right arch waist, schemes 1 and 3 respectively are 5.4%, 1.1%. The first through the NATM tunnel with large cross section of shield tunnel after excavation of small section of initial support axial force is bigger significantly than the first through hole after digging a hole, and the large section tunnel with double side heading method results is in maximum. Therefore, scheme 2 can maximize the tunnel lining support ability.

$$\text{ratio} = (\text{results of the selected scheme} - \text{results of scheme}\,2)/\text{scheme}\,2.$$

Table 2. Results of preliminary lining axial stress (unit: kN·m)

Construction scheme	Left hance	Vault	Right hance	Ratio/%		
				L. hance	Vault	R. hance
Scheme 1	1263.0	761.3	1239.0	−6.4	−5.5	−5.4
Scheme 2	1350.0	835.1	1310.0	–	–	–
Scheme 3	1330.6	826.5	1295.0	−1.4	−0.6	−1.1

(2) Mechanics analysis for small section tunnel in left line.

According to Table 3, scheme 2 by the construction scheme of segment lining negative moment is the smallest, scheme 3 are less than 1. The first segment lining moment through the NATM tunnel with large cross section of shield tunnel after excavation of small section of a first through hole after the excavation of large hole to be small,

Table 3. Calculation results of segment moment (unit: kN·m)

Construction scheme	Left hance	Vault	Right hance	Ratio/%		
				L. hance	Vault	R. hance
Scheme 1	7.8	10.2	8.1	290.0	251.7	326.3
Scheme 2	2.0	2.9	1.9	–	–	–
Scheme 3	4.5	3.9	5.6	125.0	34.5	194.7

the large section tunnel with double side heading method produces results are minimum. Therefore, scheme 2 is most conducive to the safety of tunnel linings.

(3) Analysis of surface displacement.

The subway is usually built in the city population and built-up area, so the subway tunnel on the ground for the uplift and subsidence have strict requirements, at the same time, pay more attention to and control of builders and managers. Choose the left, 3 measuring points right line tunnel and middle soil vertical axis and the surface intersection, as shown in Table 4 the final surface subsidence of 3 schemes in the measuring point. According to Table 2, the minimum ground surface settlement scheme, scheme 3 times, but the difference is small less than 1. In the middle of the surface settlement, 1, 3 and 2 respectively are in the 14.2% scheme, 4.7%. That is the first small hole after digging a big hole to small surface settlement first through the NATM tunnel with large cross section of shield tunnel after excavation of small section production, the surface of which adopts double side heading method and CRD method excavation hole the settlement difference is not large.

Table 4. Calculation results of ground settlement (unit: mm)

Construction scheme	Left tunnel	Mid.	Right tunnel	Ratio/%		
				L.	M.	R.
Scheme 1	9.2	12.1	15.6	16.5	14.2	13.0
Scheme 2	7.9	10.6	13.8	–	–	–
Scheme 3	8.3	11.1	14.2	5.1	4.7	2.9

6 Discussion

The tunnel is located in the loose loess strata, linking NATM tunnel scheme 1 larger disturbance to surrounding rock. It will have a larger surface deformation. Compared to that, after the excavation of small section tunnel by shield method in schemes 2 and 3 through grouting, chemical grouting and reinforcement and enhance the overall stability of surrounding rock of tunnel construction measures such as freezing, compared to the former can be more effective, more timely control of surface deformation and ensure the stability of the intermediate soils. This is the first new Austrian method for large section by shield method of small section tunnel with small spacing construction schemes 2 and 3 selection provides safety reserves and recovery time on theory and technology.

7 Conclusion

The left small section tunnel constructed by shield and the right section tunnel by double side heading method or CRD method with small spacing parallel sequence is researched. The total of 3 kinds of construction schemes on theory and technology selection and optimization analysis with the numerical simulation results of the scheme of tunnel lining and surrounding rock construction the mechanical state of the cause, draws the following conclusion:

(1) Constructed firstly by NATM tunnel with large section and small section tunnel boring shield method construction scheme has less disturbance to the surrounding rock, and the first through hole after the excavation of shield method NATM hole scheme, will be more conducive to reducing the vault crown settlement, stress control surface deformation and middle soil body.
(2) Constructed finally the shield of small section tunnel, a CRD method is less safety and reliable surrounding rock mechanics state than the use of double side heading method first large section tunnel.
(3) The tunnel is located in the loose loess strata, linking NATM tunnel scheme 1 larger disturbance to surrounding rock. It will have a larger surface deformation. Compared to that, after the excavation of small section tunnel by shield method which can through chemical grouting to enhance the overall stability of surrounding rock of tunnel construction measures such as freezing, compared to the former can be more effective, more timely control of surface deformation and ensure the stability of the intermediate soils.

References

Liu, B., Tao, L., Ding, C., et al.: Prediction for ground subsidence induce by subway tube tunneling. J. China Univ. Min. Technol. **35**(3), 356–361 (2006)

Deng, J., Zhu, H., Ding, W.: Finite element simulation of whole excavation operation of a unequal-span double-arch tunnel. Rock Soil Mech. **25**(3), 477–480 (2004)

Jiang, X.: Finite element simulation of construction method of crossover tunnel with a unequal-span double-arch. Railway Stan. Des. **10**, 103–105 (2009)

Gong, J.: Study on construction mechanical behavior of highway tunnel with large section and small spacing. Ph.D. thesis. Tongji University, Shanghai (2004)

Kong, X., Xia, C., Qiu, Y., et al.: Study on construction mechanical behavior of parallel small spacing metro tunnels excavated by shield and cross diaphragm (CRD) method in loess region. Rock Soil Mech. **32**(2), 516–524 (2011)

Mu, C., Li, B.: Organic matter content and mechanical properties of soft soil effect analysis. Hydrogeol. Eng. Geol. **3**, 42–46 (2008)

Qin, B., Cen, H., Liu, Y., Wang, J.: Gao miaozi swell-shrink deformation properties of bentonite and its influencing factors of the study. Chin. J. Geotech. Eng. **30**(7), 1005–1010 (2008)

Sun, W.: Swelling the hydraulic and mechanical properties of unsaturated soil and elastoplastic constitutive model. Master's degree thesis. Tongji university, Shanghai (2009)

Wang, Z., Liu, Q.-S.: Sand compacted bentonite mixture experimental study on the expansion properties. Rock Soil Mech. Chin. J. Geotech. Eng. **21**(4), 331–334 (2000)

Reducing the Tunnelling Effect on Adjacent Pile Foundations

Ahmed N. EL-Attar(✉)

Civil Engineering Department, Higher Technology Institute,
Tenth of Ramadan City, Egypt
Ahmed_civil_hti@yahoo.com

Abstract. Tunnel excavation may induce adverse effects on nearby existing structures/foundations in urban areas. Jet grouting is used as solution in the influenced area near or around the tunnel vicinity to reduce the induced stress and deformation of tunnelling process on nearby deep foundation. The objective of the present study is to numerically evaluate the nature of interaction between the employed tunnelling technologies; the nearby structures founded on deep foundation systems and grouted soil using 3D finite element idealization. Back analysis of The stability of El-Attabe Garage building due to the execution of the Greater Cairo Metro Line 3-Phase 1 is conducted in this paper to evaluate the capability of the proposed numerical model. Also, in the present study, a parametric study is conducted using a 3D finite element code to investigate the effect of the different grouting shapes such as jet grouting wall placed at or below tunnel springline, to reduce the tunnelling effect on adjacent pile. In addition, another shape of circular annuals grouting is investigated in this paper. Accordingly, the upper half and full annulus of soil region around the tunnel is grouted before tunneling process to enhance the soil properties and decrease the effect of tunneling. These grouting annuls have a same thickness and length as that of grouting wall used behind the pile. All of these schemes of grouting shapes are used to study their effect on reducing the pile response due to tunnelling process.

Keywords: Cairo Metro Line 3 · Finite element · Foundation · Grouting · Piles · Three-dimensional modelling · Tunnelling

1 Introduction

In an urban environment, due to a lack of available space, a lot of underground constructions such as tunnelling are increasing. Underground tunnelling will inevitably induce varying degrees of ground movement towards the excavation resulting in ground surface settlement. One of the important issues of tunnelling in urban areas is the assessment of likely impact on adjacent buildings and other environmental hazards. Peck (1969) developed a method to determine surface settlement and lining stresses induced by tunnelling based on field measurements and empirical data. Analytical and empirical solutions of similar problems were also developed by several researchers to calculate the surface settlement and lining stresses.

S. Agaiby and P. Grasso (eds.), *Engineering Challenges for Sustainable Underground Use*,
Sustainable Civil Infrastructures, DOI 10.1007/978-3-319-61636-0_13

One of the methods used to decrease tunnelling effect on adjacent building is by increasing soil stiffness specially near or around the tunnel vicinity. The soil stiffness improvement includes the installation of jet grout blocks.

According to Essler and Yoshida (2004), the earliest patent for Jet-Grouting was applied for in England in the 1950s; however, its real practical development took place in Japan and it was applied first to create thin cut-off walls. In the early 1970s, rotating Jet-Grouting emerged in Japan and in the mid-1970s, Jet-Grouting was exported to Europe and since then, it has become popular worldwide specially in tunnelling projects (Essler and Yoshida 2004). During the first decade of the new millennium and also nowadays, the utilised equipment was constantly improved and played a significant role in the evolvement of the technique. Thus, higher pressures and flow rates were achieved; recently, the parameters of the nozzle's upstream, such as the length and shape of the passage have been optimized in order to generate a more focused jet, in addition to the use of air (Yoshida 2010).

Jet grouting versatility and flexibility together with its applications, in almost every soil formation, makes it a perfect solution for difficult geotechnical problems. It is effective in open field as well as in confined space with limited headroom, since the column diameter does not correspond to the size of the rig. In the last decade, the unique features of this technology were used in all high profile tunnelling projects in order to facilitate the building process and to improve the level of safety and efficiency. Lee et al. (2012) carry out 3D finite element analyses (FEA) to investigate the behaviour of piles in a large pile group and associated building in to investigate the effectiveness of grouting around the tunnel in mitigating pile/building movements.

In this paper is divided into two parts; in the first part the case history of greater Cairo Metro is idealized in the analysis to evaluate the proposed 3D finite element in modelling of ground-tunnelling-piles interaction while in the second part, The closer pile cap of Garage EL-Attaba to the tunnel route of Cairo Metro (F1) is utilized in the analysis to evaluate the efficiency of grouted shapes in improving the behaviour of piles during tunnelling process.

2 Garage EL-Attaba (Greater Cairo Metro Line 3, Phase 1) Case Histroy

Greater Cairo Metro is considered one of the most important national projects executed during the second half of the twentieth century. The project consists of three lines linking the capital districts with the centre of the city. Greater Cairo Metro Line-3 is 33 km-length extends from Imbaba to Cairo Airport and planned to construct in four phases. The tunnel executed utilizing a slurry shield (TBM) of 9.55 m external diameter with tail-skin grouting. The tunnel has an internal diameter of 8.35 m, external diameter of 9.15 m and a precast segmental lining thickness of 0.4 m.

The effects of construction of Greater Cairo Metro, Line 3, Phase-1 (extended from Attaba to Abassia) on the nearby Garage El-Attaba building (supported on pile foundation) is numerically analysed optimizing 3D finite element modelling. Garage El-Attaba is 45.55 m-width and 77.04 m-length, and the tunnel is running making an angle of about 42° from the building with a nearest distance to the tunnel centreline of about 6.45 m. The building is skeleton-type comprises eight floors founded on piles 0.60 m-diameter & 20 m-length).

The preliminary sensitivity analysis of the present case study indicates that, the tunnelling process only affects the enclosed pile capes (F1, F2, F3, and F4), Fig. 1,

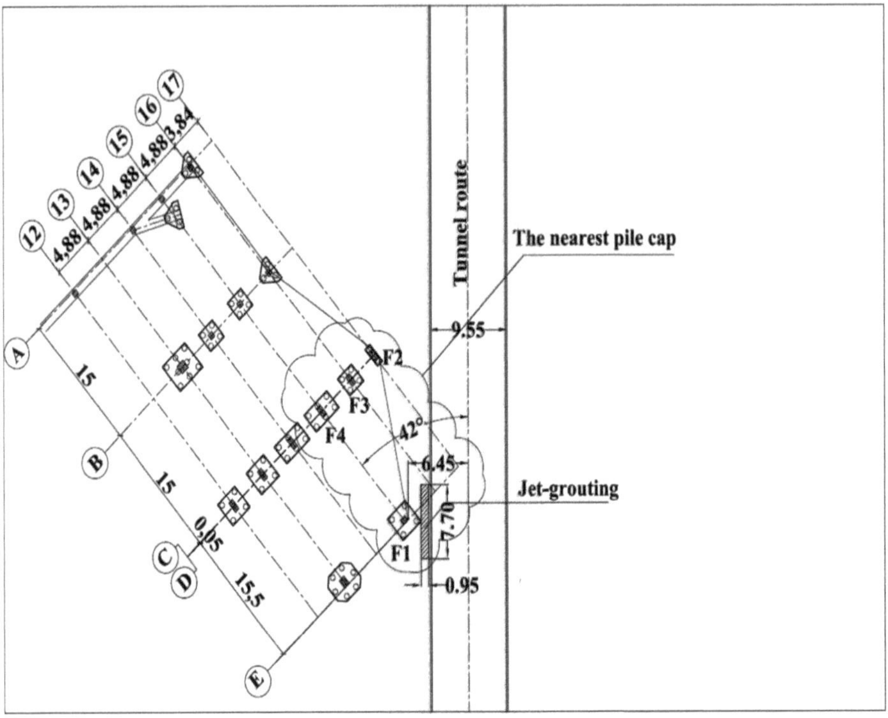

Fig. 1. Cross-section and plan views of the closer pile caps relative to the tunnel route.

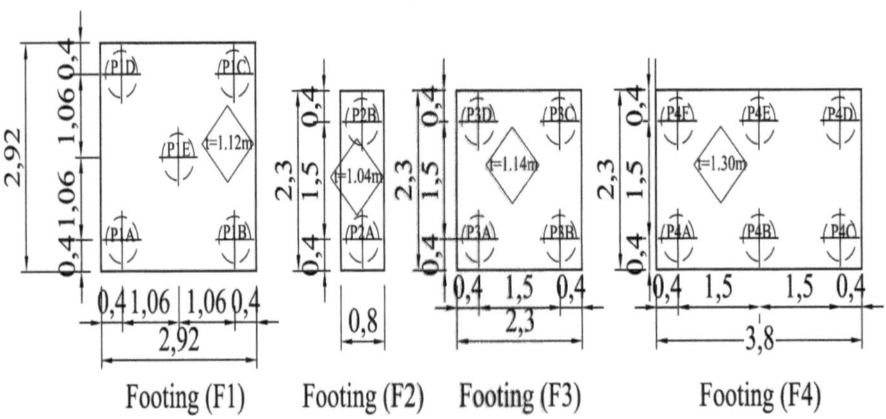

Fig. 2. The pile caps dimensions, (after Mocarthy Brothers Company 1984).

while the rest surrounding pile caps indicate minor effect. Figure 2 shows the dimension of each pile caps as well as the spacing between piles.

3 Jet Grouting– A Solution to Problems of Tunnelling Effect on Adjacent Building

The primary goal of grouting is to increase the strength of the native soil and reducing their permeability. Permeation grouting is a ground modification technique that fills pore spaces in soil with grout. Permeation grouting is most often used to decrease permeability, limit settlement by increasing the cohesion of the soil, and thus reduce ground loss during tunnelling. Various materials are used for this type of grouting, including ordinary Portland cement, microfine cement, bentonite, and most often sodium silicate. There are two basic steps in the jet grouting process - drilling and grouting. There are two basic steps in the jet grouting process - **drilling and grouting**. In this case history a compensation of cement, bentonite, and water is used in **drilling**

Table 1. The grouting mix used to change soil properties

Cement (kg)	Water (kg)	Bentonite (kg)
750	750	15

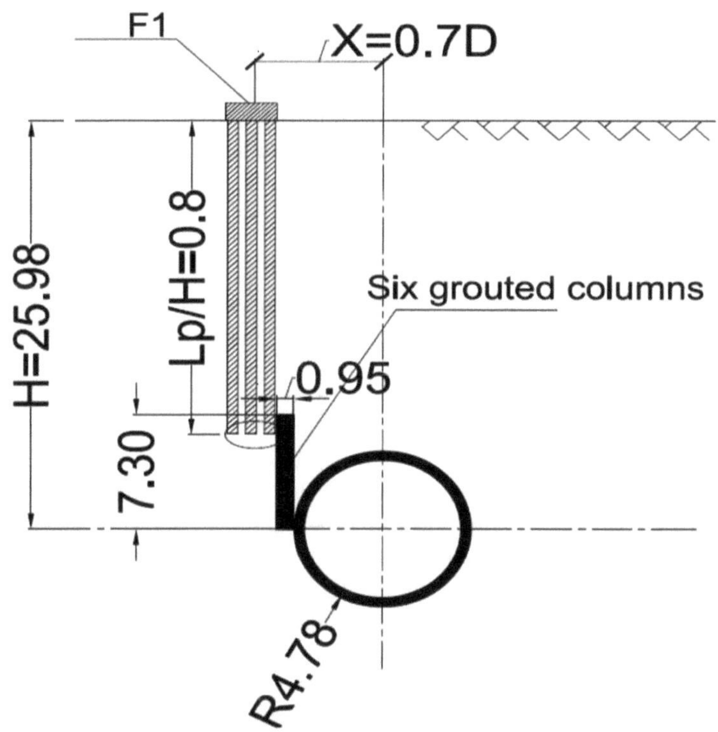

Fig. 3. The grouting columns location relative to tunnel vicinity used in the case history.

to fill any encountered cavities (if any). The grout mix used in **drilling** consists of 350 kg cement, 875:900 kg water, and 25:30 kg bentonite. After drilling of each hole, a bundle of five grout pipes was installed into the borehole filled with cement. Grout started the next day with the following mix shown in Table 1.

Accordingly, the soil properties in region near the tunnel vicinity were enhanced by using 6-tangent-jet-grouted columns, which have an overall horizontal distance of 7.7 m. The size of each column is 1.3 m, and a depth of 7.3 m which is calculated based on the anticipated strength of the jet-grout improved soils, the width between the tunnel and adjacent pile and length of the nearest piled footings relative to the tunnel route. The jet-grouted columns was performed as shown in Fig. 3. The grouting changes the soil properties in region near the tunnel vicinity as shown in Table 2. After the jet-grouted wall was constructed, the tunnel excavation starts.

Table 2. The estimated subsurface geotechnical parameters.

Layer	Thick. (m)	γ (kN/m^2)	$\varphi°$	C (kN/m^2)	E (MN/m^2)	ν	Ψ
Made ground	6.06	17.0	27	0.0	4	0.3	0°
Silty clay	3.58	19.5	$\varphi_u = 0°$	$C_u = 90$	$E_u = 22$	$\nu_u = 0.49$	0°
Upper sand	13.81	19.5	36°	-	40	0.3	6°
Sand & Gravel	1.15	20	41°	-	100	0.3	11°
Middle sand	1.38	19.5	38	-	70	0.3	8°
Lower sand	Ext.	19.5	38°	-	120	0.3	8°
Grouted soil	Var.	22	52	390	1500	0.15	22°

4 Numerical Modelling

The finite element mesh used to simulate the reference case is shown in Fig. 4. Numerical modelling is idealized by means of the finite element program ABAQUS, (Hibbitt and Sorensen Inc. 2011). The model boundary condition is determined after conducting sensitivity analysis. The analysis of the tunnelling–structure interaction is performed on two steps. The first is concerned with the determination of initial stresses in the soil mass prior to the tunnel construction, considering the self-weight of the soil, piles, and pile cap. The second step deals with the simulation of the tunnelling advancement in presence of the pile cap. The 3D finite element modelling of the ground-tunnelling-piling interaction is visualized taking into consideration the details of tunnelling activities and the rate of tunnel advance. Advancement of TBM was simulated simultaneously as follows; (i) applying the excess face pressure, (ii) removal of soil elements and activation of shield, and (iii) activation of lining elements and grouting elements from the liquid to hardening state, (Pang et al. 2005). The excavation rate activity is simulated for 3 m/step of the tunnel advancement. The rate of the tunnel

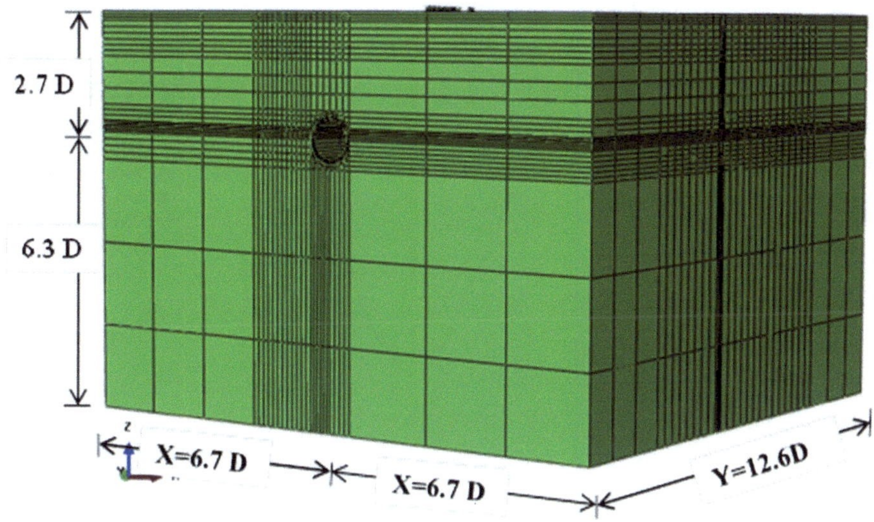

Fig. 4. Finite element modeling of Garage EL-Attabe case study.

advancement (9 m/day) is updated in the finite element modelling to account for face pressure-excavation-lining erection-tail grouting-grouting setting consequences.

The elastic-plastic Mohr-coulomb failure criterion is adapted in analysis to model the soil behaviour. The structural properties of the lining and the shield are assigned as linear elastic material. The shield is modelled as solid elements having (E = 200 GPa and ν = 0.25). Also, the over-cutting is modelled using solid elements having (E = 1000 kPa and ν = 0.2) and representing the grout properties in the liquid state, while the hardened grout is assigned to (E = 50 MPa and strength of about 0.5 kPa after 28 days.

5 The Numerical Model Result

Figure 5 shows the surficial vertical settlement trough at the location of elevation reference point according to the two cases of ungrouted (original case as in field) and grouted soil near the tunnel vicinity. On the same figure the distribution of settlement trough profile for the green field conditions (no assignment of the structure) is allocated and the building field settlement compiled during tunnel crossing at the point of elevation reference point (E.R.P.A). The results indicate that the computed results in case of ungrouted soil express relative conformity with building field measurements compiled during construction consequence. This agreement between the two results (field measurements and computed data) indicates the capability of the proposed numerical model in simulation.

On other hand, the exiting of grouted block reduce the surficial settlement under the building more than that developed in the original case by about 28.5%.

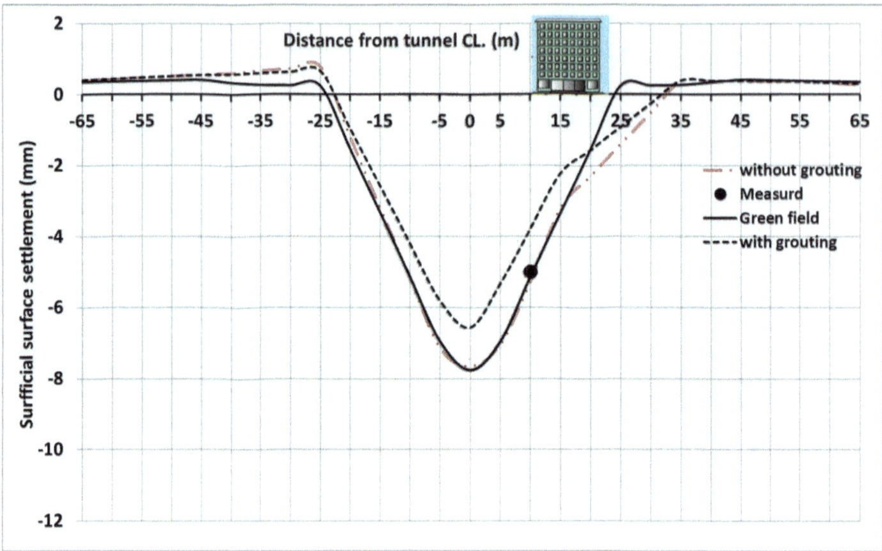

Fig. 5. The computed vertical settlement trough at the location of the elevation reference point due to grouted and ungrouted case.

6 Parametric Study

The parametric study of the present work is focused on the effect of six jet grouted columns on changing the pile behaviour. The same numerical model used in the verification case is used to evaluate the capability of the grouting in reducing tunnelling effect on adjacent piled footing.

Due to space limitations, only results for the most critical pile (closer pile (P1B) will be disused in detail. The analysis are carried out on the two pile length (Lp/H = 0.8, Lp/H = 1.4) located at horizontal offset from the tunnel route equal 0.7D. The choice of pile positions is according to Eid et al. (2015). Where, The short pile (Lp/H = 0.8) is located in zone (A) which is characterized by settlement greater than greenfield and the maximum tensile force while the long pile is subjected to maximum drag force and minor settlement (Zone (C)).

In total, three 3D analyses have been carried out as summarised in Figs. 6 and 7. Analysis (1) is the baseline analysis with no grouting. Analysis (2) investigates the effect of the Jet grout columns of 0.95 m thick and their toe located above the tunnel crown. Analysis (3) models a 0.95 m thick and their toe located below the tunnel crown, while analysis (4) and (5) investigates the effect of upper half and full thick grouted annulus around the tunnel respectively.

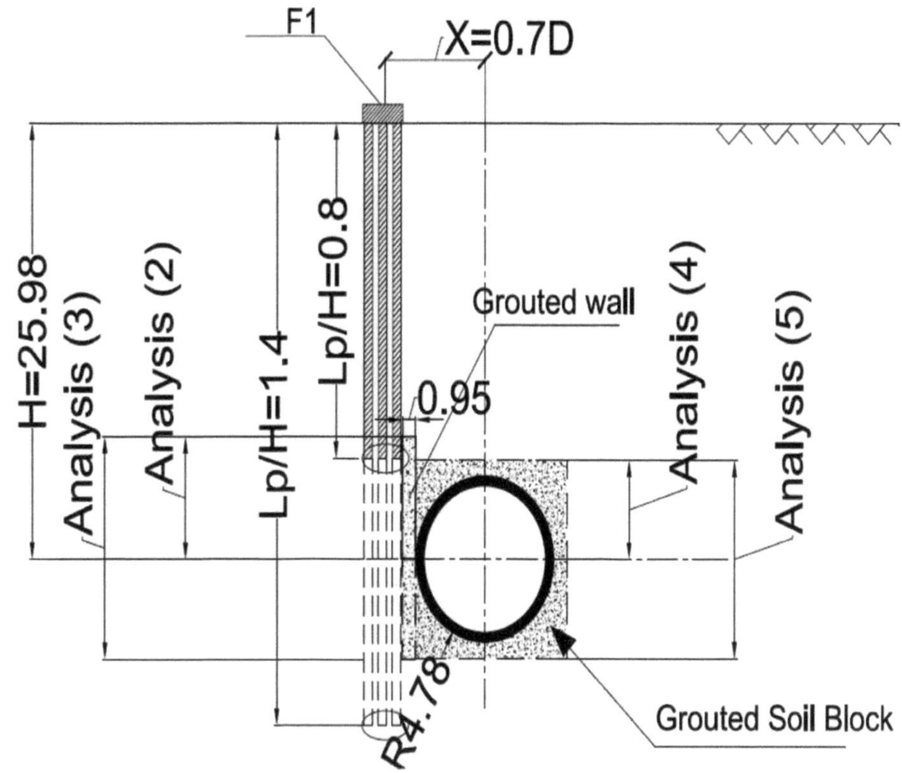

Fig. 6. The locations of pile tips, grouted columns, full annuls, and upper half grout relative to tunnel axis.

7 Results

Results for the parametric study Analyses (2), (3), (4), and (5) are compared to the baseline Analysis (1) in terms of vertical settlement, lateral displacement, and axial force.

7.1 Pile Vertical Displacement

Figures 8 and 9 show the pile (1B) vertical displacement during the tunnel advancement between the five analysis types according to the two pile configuration. It can be seen that in case of short pile the six jet grouted columns (Analysis 2 & 3) reduces the pile settlement more than that developed in case of full or upper half circular grout around the tunnel. However, the greater reduction in short pile settlement is about 60% in case of analysis (3), while The half and full grouted annulus reduce the maximum pile settlement by about 11% and 40% that developed in case of no grout respectively.

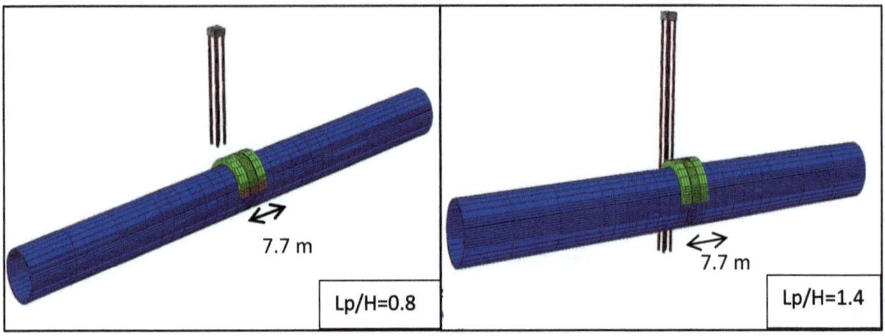

[a] Analysis (4): Upper half annuls grout

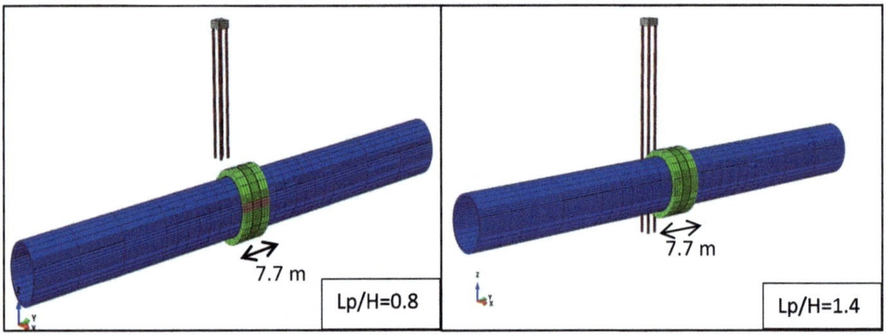

[b] Analysis (5): Full annuls grout

Fig. 7. The confining grouted annulus around the tunnel [a] Analysis (4): Upper half annuls grout [b] Analysis (5): Full annuls grout

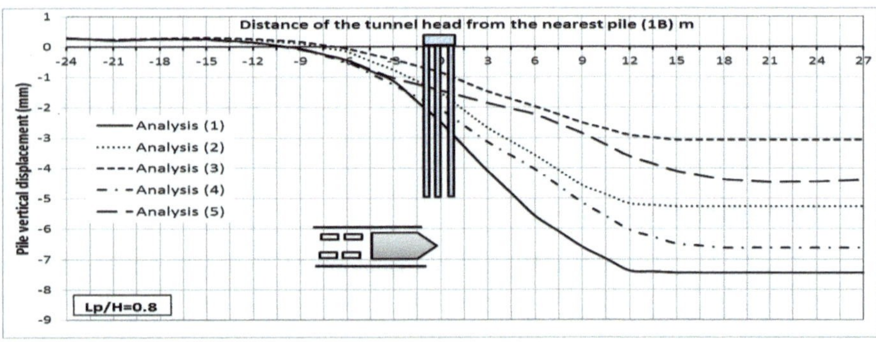

Fig. 8. The vertical displacement of front pile due to tunnel advancement for pile length ratio (Lp/H = 0.8)

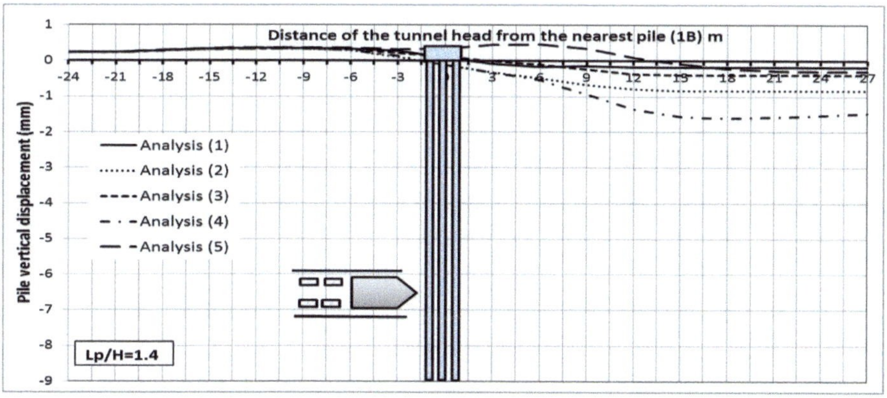

Fig. 9. The vertical displacement of front pile due to tunnel advancement for pile length ratio (Lp/H = 1.4)

Comparing the results of vertical long pile displacements associated with tunnelling indicated the efficiency of full grouting annulus around the tunnel in reducing the pile displacement and vice-versa in the other cases. This reduction is great and reaches about 70% of that developed in case of analysis (1).

7.2 Lateral Displacement

Figure 10, illustrates the lateral displacement of the nearest pile (1B) due to the five cases of analysis types for the two pile length configuration when the TBM. is behind the face by 24 m. Generally for both pile configurations, it can be noticed that as the jet grouted columns length increases the maximum lateral displacement of the short pile decreases more than that developed in the other cases. It can noticed for short pile that the existence of grouted soil reduces the lateral displacement near the head and tip of short pile, while along the pile depth the lateral displacement is the same along the pile shaft. Otherwise, in case of confining grouting around the tunnel decreases the lateral displacement at the tip of short pile more than that developed in the other cases. However, in case of long pile all grouting shapes reduce the lateral displacement specially in the region facing the tunnelling opening. To conclude, the results indicates the capability of grouting placed below the springline of the tunnel in reducing the lateral displacement which is great (analysis 5) and reaches about 60% in case of short pile while in case of longer pile the reduction is about 28% (analysis 3). The reason for this observation is attributed to the location of jet grouted columns in the front of tunnel vicinity where the ground movement is great.

7.3 Axial Force

Figure 11 shows the induced maximum axial force along the nearest pile shaft for the two pile lengths (Lp/H = 0.8), and (Lp/H = 1.4). By looking closely into the induced

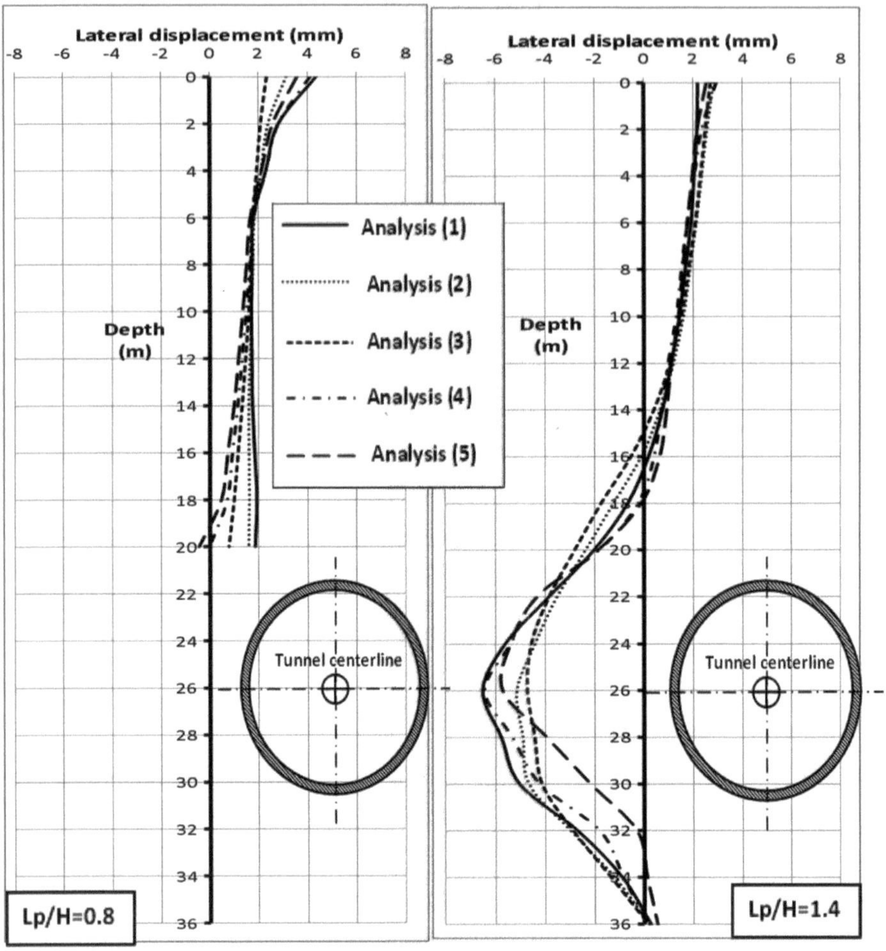

Fig. 10. The lateral displacement of front pile for pile length ratio (Lp/H = 0.8), and (Lp/H = 1.4).

axial forces of short pile it can be noticed the effectiveness of the grouted annulus schemes specially in case of analysis (2, 3, and 5) in reducing pile settlements and preventing the pile toe from experiencing tensile forces. This result is in a good agreement with the results obtained by Lee et al. (2012). Also, it can be seen that the full grouting annulus (Analysis 5) around the tunnel shaft reduces the induced axial force more than that induced in other cases which implies the efficiency of jet grouting columns at that position. On the other hand, it can be outlined that for long pile, the grouting schemes (Analysis 3, and 5) predict compressive axial force throughout the pile length less than the allowable axial force (1200 kN) (according to Mocarthy Brothers Company, 1984, the consultant of Garage EL-Attaba) by about 10% and 20% respectively. Contrary to the axial force induced in analyses (3) and (5), the axial force near the lower part of the pile increased in analysis (2) and (4) more than that induced

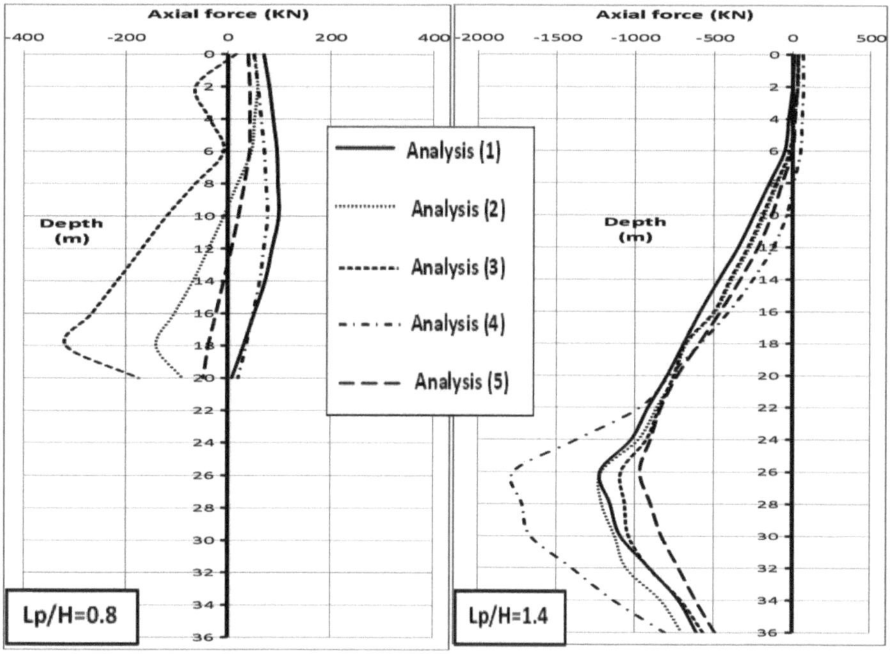

Fig. 11. The induced axial force of front pile for pile length ratio (Lp/H = 0.8), and (Lp/H = 1.4).

in analysis (1). The reason for this observation is probably attributed to the increase in the stress in the soil around the pile due to grouting. The maximum axial compressive force achieved along the pile shaft in case of analysis (1), (2), (4) is about 1211 kN, 1220 kN, and 1770.85 kN respectively which is larger than the allowable compressive force. Hence, there is high risk of developing cracks in the pile.

8 Conclusion

- Back analysis of the case history shows that there is good agreement between the measured and computed results which resulted from the proposed 3D numerical modelling.
- Applying the proposed 3D numerical modelling of the ground-tunnelling-piled foundation interaction on the case history of large tunnel projects constructed closer or nearby piled foundation indicate the capability of the proposed numerical model to update the ground deformations associated with tunnelling.
- All grouting shapes reduce the maximum short pile settlement during tunnelling process specially in case of full annulus grouting and long jet grouting column (placed below tunnel spring line). This reduction in settlement preventing the short pile of experiences tensile force near the pile toe. This result suggests that grouting

work carried out close to the source of ground movements around/in the tunnel area is effective in mitigating tunnelling-induced ground movements.

- Full annulus grouting around the lining and the long grouting wall placed below tunnel springline proved to be a very salutary process to minimize the long pile vertical displacement and hence reducing the induced axial force resulting from tunnelling. Otherwise, the other cases of analysis increase the pile displacement. Accordingly, the pile compressive axial force exceeds the allowable axial force. So it is recommended to use full annulus grouting to minimize the tunnelling effects in case of long pile which tip located in zone (c).
- All grouting schemes reduce the short pile lateral displacement either near the pile toe in case rounded annuals grouting or near the pile head in case of long and short jet grouting walls.
- For piles located in the zone (C) (long pile), all grouting schemes reduces the pile lateral deflection in the region facing the tunnel opening specially in case of long jet grouting wall.

References

Eid, M.M., Ahmed, A.A., Hefny, A.M., El-Attar, A.N.: Study on some factors affecting Ground-Pile-Tunnelling interaction. J. Am. Sci. **11**(6), 263–274 (2015)

Essler, R., Yoshida, H.: Jet Grouting. In: Moseley, M. P., Kirsch, K. (eds) (2nd Edition, London), Ground Improvement. Spon Press, New York, pp. 131–168 (2004). Chapter 5

Hibbitt, K., Sorensen, Inc.: ABAQUS User's manual, Version 6.11 (2011)

Lee, S.W., Choy, C.K.M, Tse, S.C.: 3D Numerical modelling of tunnelling intersecting piles. In: Viggiani, G., (ed) Geotechnical Aspects of Underground Construction in Soft Ground. Taylor & Francis Group, London, pp. 919–925 (2012)

Mocarthy Brothers Company: A division of design and management, consultant of Attaba Parking Garage, Cairo, Egypt (1984). National Authority for tunnel (NAT): Tunnel from Attaba to Geish shaft monitoring measurements, contract N° 49/Metro, phase 1, Greater Cairo Metro, pp. 270–271 (2010)

Pang, C.H., Yong, K.Y., Chow, Y.K.: Three-dimensional numerical simulation of tunnel advancement on adjacent pile foundation. In: Erdem, Y., Solak, T. (eds) Analysis of the Past and Lessons for the Future. National University of Singapore, Singapore (2005). ISBN 04 1537 452 9

Peck, R.B.: Deep excavations and tunnelling in soft ground. In: Proceedings of the 7th International Conference on Soil Mechanics and Foundation Engineering, pp. 225–290, State of the art volume. SociedadMexicnan de Mecanica de Suelos, A. C (1969)

Yoshida, H.: The progress of Jet Grouting in the last 10 years in Japanese Market. In: Proceedings of 35th Annual Conference on Deep Foundations 2010, Deep Foundation Institute, pp. 157–167, Hollywood, California (2010)

Numerical Modeling by Plaxis Software (3D), the Effect of Digging a Tunnel on the Behavior of the Ground and Overlying Structures

Case: Subway of Algiers (Algeria)

Djamel Eddine Boudjellal(✉), Abdellah Hafsaoui, and Talhi Korichi

UBMA: Badji Mokhtar University, BP12, 23000 Annaba, Algeria
djameljimi28@gmail.com,
abdellah.hafsaoui@univ-annaba.dz,
talhikorichi@yahoo.fr

Abstract. Analysis of interactions between underground construction projects and both ground and surface structures is an increasingly strong issue in the geotechnical field, especially in urban areas during an ongoing underground construction project or an excavation project near existing buildings.

Tunneling operations at a shallow depth in an urban site (e.g.: Algiers subway) induces, in extreme cases, ground movements that are capable of causing deformations and considerable damages to the nearby above ground structures. This work addresses the effect of the sequential excavation tunnel in by the Austrian method (NATM) at a shallow depth in an urban area, through a numerical model using Plaxis software (3D). The results of this model will be compared to those obtained by an In-Situ measurement comparing, in order to validate the numerical model and to make it easier to predict land deformations in an excavation zone.

Tunneling by the Austrian method (NATM) at a shallow depth in an urban area, through a numerical model using Plaxis software (3D). The results of this model will be compared to those obtained by an In-Situ measurement comparing, in order to validate the numerical model and to make it easier to predict land deformations in an excavation zone.

Keywords: Tunnel · Subway · NATM · Plaxis 3D · Sequential excavation

1 Introduction

One of the major problems related to these works is consisted by the movements of the ground induced by work especially with low depth in an urban site. (Case: Subway of Algiers), induced ground movements, likely to cause deformations, and even significant damage to structures on the surface in extreme cases. This problem is all the more important in the presence of very poor geological formations such as shale and the argillite strongly affected, the settlement in this case develops in the course of time, Sometimes over long periods of time after the end of the work.

S. Agaiby and P. Grasso (eds.), *Engineering Challenges for Sustainable Underground Use*, Sustainable Civil Infrastructures, DOI 10.1007/978-3-319-61636-0_14

The engineer, therefore, is faced to two problems; preserving the stability of the structure and controlling the deformation of the massif. Thus, various techniques can be associated Such as the use of the vault umbrella or the strengthening of the front size by bolting, In order to maintain the front from collapsing and minimize the settlements on the surface.

The study of the tunnel balance and stability is of a major interest. The movements which are likely to occur at the level of the front and on the surface have helped to introduce recent achieving and strengthening methods so that we can control them. The aim of this work is to study numerically using Plaxis software (3D) the stability of a tunnel section.

(Section D9 Ain Naaja -Baraki) using a three-dimensional calculation code has been used to better visualize the movements at the front and also to make an excavation by step.

2 Presentation of the Project Ain Naadja-Baraki

2.1 Introduction

Due to the strong demand for transport, The Department of Transport has decided to carry out the execution of the preliminary study, Preliminary Project Summary (P.P.S.) and Detailed Preliminary Project (D.P.P.) of three extensions of the Algiers subway

- Extension El Harrach Centre- Bab Ezzouar (lot 1)
- Extension Ain Naadja - Baraki (lot 2)
- Extension Place of The Martyrs - Bab el-Oued - Chevalley (lot 3)

The Subway Company of Algiers assigns to the Grouping IDOM- EM, IDOM being the leader of the grouping, the achievement of the preliminary study, Preliminary Project Summary (P.P.S.) And Detailed Preliminary Project (D.P.P.) of the extension of the Subway of Algiers, Lot 2: Ain Naadja- Baraki. The proposed route for the Preliminary Project Summary (P.P.S.) starts at the level of the Centre of Ain Naadja and ends at Baraki, Its length is approximately 6.16 km with a minimum ray of 350 m and a maximum of 5500 m and with six (6) stations (Fig. 1).

Once the plot is analyzed in more detail in the Preliminary Project Summary (P.P.S.), the plot is made at the level of the Detailed Preliminary Project (D.P.P.) of the section between the urban cores of Ain Naadja, till the exit to the interurban zone of the plot.

The projected route for the Phase 1 of the Detailed Preliminary Project (D.P.P.) begins at the entrance of Ain Naadja and ends at its exit passing by the interurban zone, its length is approximately 2.33 km with a minimum ray of 350 m and a maximum of 1000 m, passing by two (2) projected stations.

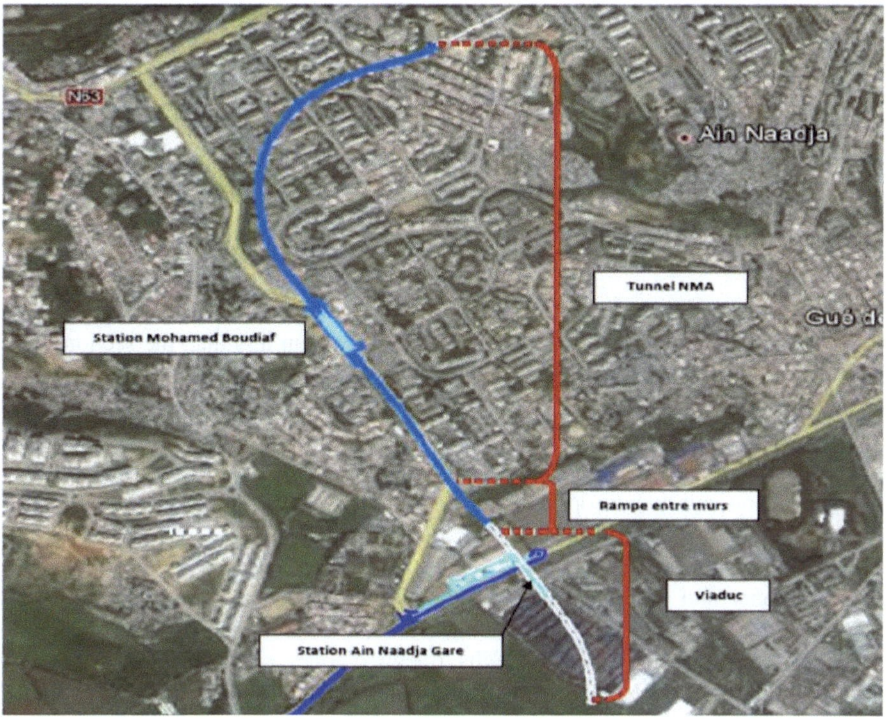

Fig. 1. Description of the route Ain Naadja-Baraki

2.2 Description of the Route

The route of the portion of the subway of Algiers, subject of our study, is carried out by the urban core of Ain Naadja most of which is underground through a NMA tunnel. Once the tunnel route of Mohamed Boudiaf Avenue finished, the tunnel passes to a section in the tunnel framework to exit on the surface; a viaduct of the exit of Ain Naadja is projected. The projected route begins at the last ventilation well (VW6) of the extension C, currently in execution: Hai El Badr - El Harrach and Hai El Badr - Ain Naadja, extending the straight alignment of the prior segment and with a section in the NMA tunnel composed of a vault of 3 spokes.

The presence of buildings on the surface, in addition to the low depth of the previous segment height, already built makes necessary a strong decrease of rating.

The height obtained on the vault of the tunnel between the P.K. 0+100 and P.K. 0+200 is between 6 and 7 m. Due to the presence of buildings, it will be necessary to include some geotechnical treatments on the ground for the protection of these buildings.

3 Study and Recognition of the Soil

3.1 Geological and Geotechnical Study

3.1.1 Introduction [9]

Each tunnel has its own geometry, location and natural conditions. Despite this, the achievement of a tunnel remains a succession of repetitive operations in all projects that the responsible of the project follow: The recognition of the environment, digging, retaining and coating.

3.1.2 Reconnaissance Galleries [2]

A gallery of recognition is a gallery of reduced dimension less than the section of the definitive work in which it is registered. It allows to better assessing the difficulties of achieving of the projected work. Its objectives can be classified as follows: direct contribution to the construction; facilitation of the execution; geological and geotechnical reconnaissance of the soil allowing the analysis of the soil behaviour in reduced section and thus to optimize the construction methods.

3.1.3 Geological Reconnaissance [1]

Reconnaissance and geological studies must give the designer of the tunnel necessary elements for regional geology, the geological and tectonic history of the massif; the inventory and location of accidents (faults, fractured zones, crushed zones, …); and the description of the soil encountered depending on their petro graphic and mineralogical nature.

For these reasons, we must opt for means of geological importance such as existing geological data inventory (geological maps, previous documents and geological studies. etc.); Surface Geological Survey, the geophysical prospecting and surveys.

3.1.4 Hydro Geological Recognitions [1]

Recognition and hydro geological studies must be carried out at the same time as recognition and geological studies. They have to be for the purpose of determining the aquifers formations and impermeable sites; the hydraulic regime (groundwater, underground circulations…); the soil permeability and the karstic networks.

This information is used to define the nature of the goings possible water along the route; their load, their flow, the chemical and mineralogical composition of the waters and treatment options (Fold, drainage, sealing…). Among the means of hydro geological recognition to be established, inventory of existing data; hydro geological investigation of surface (water sources, wells); electrical resistivity method for the research of the groundwater and surveys (piezometric level, the permeability…).

3.1.5 Geotechnical Recognition [1]

Recognition and geotechnical studies complement those carried out by the geology and hydrogeology. They must allow to clarify the description of the land crossed, their state of in situ constraints and predict their behaviour during the execution and operation of the structure and to define the methods of execution, the sizing of the works, the special precautions and/or the treatments to be considered during the construction to complete the classifications. The means of geotechnical recognitions to implement are: Inventory of existing data; the surveys to carry out the measures in situ and laboratory tests.

The most widely used laboratory tests for the recognition of structures Are tests for the determination of physical properties (density, porosity, void indices, Atterberg limits, particle size and sedimentometry, mineralogy, sound speed) and mechanical (box-shear tests, tri axial tests, oenometric tests, swelling tests, simple compression, Franklin and Brazilian tests, …).

3.1.6 Geological and Geotechnical Reconnaissance of the Present Case (km-0+000 and km-0+208.8) [3]

The region of Algiers can be represented in simplistic form as a primary metamorphic dome, designated by massif of Algiers which is bordered by of the sedimentary in tertiary and quaternary formations. The route concerned by this study is located in the territory of Gué De Constantine and Baraki, around 10 km from the centre of the city of Algiers. The area pressing an anticlinorium at the level of Ain Naadja and a flat area at the level of Oued El Harrach and Baraki.

The lithological unit crossed by the planned tunnel route can be reduced to two main units which overlap more or less horizontally. The upper unit consists of quaternary sediment predominant of cohesive character.

This layer contains a large number of scree. It may be that there are anthropogenic changes (fills).

The following unit consists of clastic sediments of sandy type to clayey soils of a large heterogeneity, gathered under the term molasse.

The Formations of Quaternary [3]

The recent deposits (RX):

The upper unit, sometimes of a thickness of a few meters, usually not consolidated. Of more or less cohesive character. It is rich in the scree and can be outcome of anthropogenic changes, in the same way that it can contain construction materials debris.

A group of unique characteristics values (RX), characterized by reduced parameters has been specifically defined for this area heavily revised.

The Quaternary clay (QC):

A very cohesive clayey facies, It is characterized by a predominance of sandy and loamy clays.

The sandy quaternary (SQ):

A facies of silty sands little cohesive and in a larger part slightly subordinate clay.

The marl quaternary (MQ):

The lowest facies of the quaternary consists of gray marl very cohesive and slightly sandy. They form the lower limit of the quaternary and separate the Quaternary sediments from tertiary sediments.

The Tertiary Formations

The Tertiary clay (TC):

A mixture of clay layers sandy and gravelly. Thin sandstones layers are intercalated repeatedly and in an irregular way in the molasse sediments.

The Sandy Tertiary (ST):

This facies is a mixture of silty sands and clayey layers a bit cohesive. Thin sandstones layers are intercalated repeatedly and in an irregular way in the molasse sediments.

Achievement of Carrot Surveys

The works of surveys have been conducted by TREVI Italian Company. The recovery of the carrot is quite satisfactory exceeding the 90%.

The analyzes selected for the study of tunnel are as follows:

- 32 Carrot surveys;
- 06 Piezometric surveys;
- 24 Pressiometric surveys.

In Situ Tests [3, 7]

Pressiometric Tests. The Pressiometric tests are carried out with a probe introduced into the drilling done by a destructive with the drill LONGYEAR. The test is to determine the pressure limit (PL) of the field, the pressiometric module (E) of the soil in question and the creep pressure (CP). The E/PL report allows assessing the state of consolidation of the field.

Dynamic Penetration Tests. The dynamic penetration Tests consists of pushing the tip of a rigid probe into the soil and count the number of strokes for each sinking of 1.0 m. It allows to know:

- The succession of layers;
- The homogeneity of a layer;
- The position of a known resistant layer.

The number of tests carried out is 10.

The STP tests are performed in each carrot survey in terms of facies change where each 3 m consisted of inserting a point of 20 cm in length and with the help of a hammer. The number of strokes for each depth of 20 cm is counted in order to obtain the consistency of the ground. The tests are carried out on the whole carrot polls in total of 243 tests (The results of tests. See the Table 1).

Table 1. The characteristics of the soil of after the results of SPT

Soil Type	RX	QM	TS
Module E [MN/m^2] (SPT)	6	30	80
ES module [MN/m^2] (SPT)	25	60	135
Cohesion cu [kN/m^2] (SPT)	5	30	15
Angle of friction [°] (SPT)	28	27.5	33

Laboratory Tests [3]. The tests of analyzes are carried out on the samples collected during the coring and are intended to determine the geotechnical characteristics of the crossed layers.

(The results of tests see the Table 2).

Table 2. The characteristics of the soil of after the laboratory results

Soil type	RX	QM	TS
Wet density [kN/m^3] (Laboratory)	22.8	24	25.2
Cohesion c' [kN/m^2] (Laboratory)	5	30	15
Cohesion cu [kN/m^2] (Laboratory)	15	150	30
Angle of friction [°] (Laboratory)	28	27.5	33

3.1.7 Results of Geological and Geotechnical Reconnaissance [3]

The results of the geological and geotechnical reconnaissance executed at the level of Section D9 is the geological longitudinal cut which shows the geological and hydro geological conditions in view of a model of a simplified and generalized basement in view of the distribution of zones, the soil types of with comparable characteristics (homogeneous units) (Fig. 2).

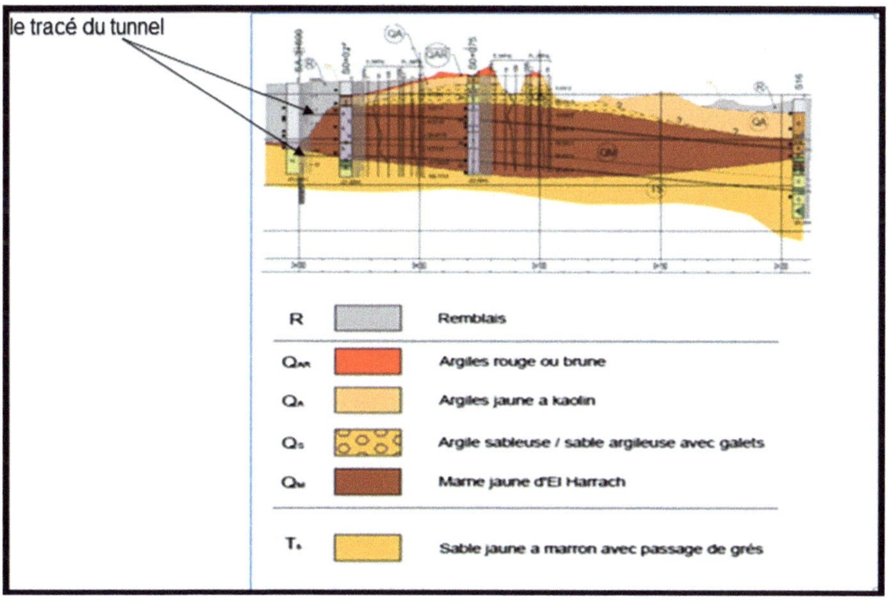

Fig. 2. Geological cut at the level of Section D9 (EMA)

The geology of the soil is summarized as follows:

A-Quaternary:

- Layer of embankments ranging from 0.40 m of thickness;
- Red clay slightly sandy, plastic, thickness of 0.30 to 1.0 m, sometimes eroded;
- Yellow Clay to kaolin- plastic sometimes silty reaching the thickness of 5.0 m, this formation is sometimes interspersed by layers of fine beige sand.
- Sand with rollers and conglomerates, layer of rounded pebbles reaching 10 cm in diameter mixed with cohesionless sand with average grain to coarse. Very unstable layers in the tunnel with a thickness of 0.50 m up to 4.0 m.
- Yellow Clay to kaolin blue or red consistent enough, plastic, to fine grains, with breaks in seals oblique and horizontal. The blocks reaching more than 3.0 m of dimensions. The thickness of this layer exceeds the 06.0 m.

B-Tertiary:

- Gray and cohesionless sand, exists only in a few surveys and forms calcareous hard gray lenses, it is molasse.
- Greenish-yellow marl with limestone debris, very consistent in thickness, exceeding the 04.0 m.

3.2 Conclusion

The geological and geotechnical studies are carried out in order to identify all problems concerning the layout of the tunnel to be crossed and for this reason; investigations are made in order to know to the maximum the details of the soil, and to be able to respond positively to the effective constraints that can encountered during the excavation.

The recommended values for Section D9, (See the Table 3).

Table 3. Characteristics values for Section D9

Soil type	RX	QM	TS
Wet density [KN/m^3]	22.8	24	25.2
Dry density [KN/m^3]	22.8	24	25.2
E module [MN/m^2]	6	30	80
Es module [MN/m^2]	25	60	135
Cohesion c’	5	30	15
Cohesion cu	15	150	30
Friction angle φ [°]	28	27.5	33
Resistance Rc [KN/m^2]	–	219	111

where:
(**RX**): The recent deposits;
(**MQ**): The marly quaternary;
(**TS**): The Sandy tertiary.

4 Techniques of Digging and Support

4.1 Introduction

The digging of the tunnel is quite complicated operation because it is related to several settings that can affect the soil or the rock, depending on the different modes of digging, the excavation of the land can be done according to the type of land encountered, the geometry of the structure, and its depth, the techniques of construction must ensure the stability of the ground while respecting certain criteria of economy, speed, and security (Fig. 3).

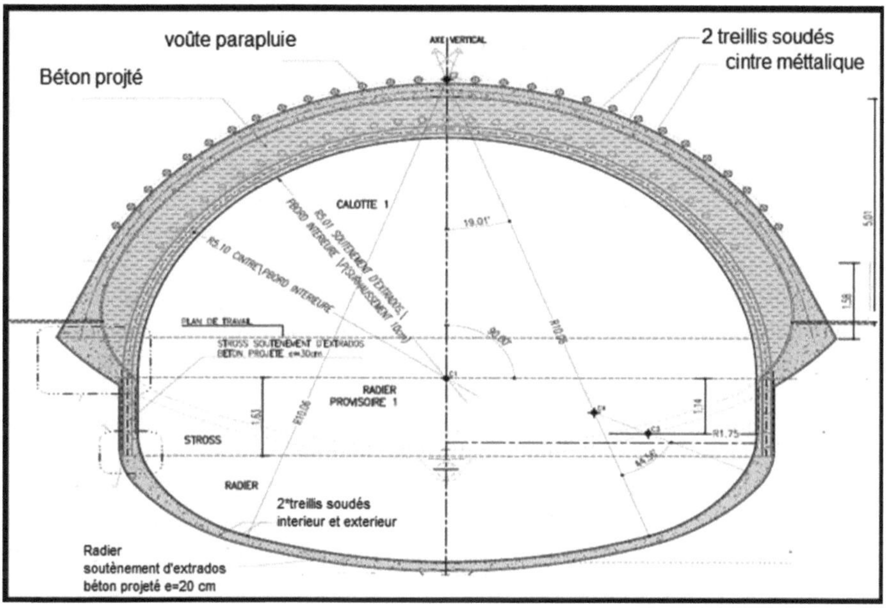

Fig. 3. Right section of retaining of the tunnel

4.2 Study of the Present Case PK 0+000 to 0+208, 85 (Section D.9) [3]

4.2.1 Description of the Execution Process

Equipments

A. Execution materials (See the Table 4).
B. Retaining Materials

- Metal Hanger HEB;
- Welded Mesh;
- Projected concrete;
- Threading;
- Anchoring.

Table 4. Equipments required for digging and retaining work

Equipment	Manufacturers
Excavator	LIEBHERR
Excavator	CAT
Wheel loader	CAT
Pump for shotcrete	MEYCO
Pump for shotcrete	CIFA
Forklift Truck	CAT
Mini Excavator	CAT
Robot of projection	CIFA

4.2.2 Method and Mode of Excavation

The excavation of the tunnel along the route of the extension C of line 1 of the subway of Algiers (Section D.9) is made by the new Austrian method (NATM), which has already proved its worth in the construction of tunnels under similar ground conditions.

The principle is to install a temporary flexible support in shotcrete and strengthen it, by welded mesh, metal hangers and anchor bolts if necessary, allowing a few ground deformations around the cavity, with an optimal combination of retaining structure, and subsequently the stability of the ground is obtained.

The choice of additional protective measures depends on the geological conditions, the initial conception is always examined during the progress of the digging on the basis of the measures of convergence and the geological conditions encountered.

The establishment of the temporary support allows us to ensure the stability of the excavation before installing a sealing membrane and to achieve the final support. To reduce the effect of constraints and initial deformations created by the opening of the cavity, it is necessary to put in place the temporary retaining as quickly as possible.

Therefore, the conventional digging is a process of cyclic execution: The repeated excavation steps are followed by the installation of an adequate support; these two steps depend on the existing geological conditions and of the behaviour of the land.

This method used mainly for the excavation of the standard equipment allowing access to the size front practically at any time, and also offers the advantage of a large flexibility in different situations.

According to the form, the dimensions of the tunnel and its height, the tunnels of segment of the extension - C- the subway of Algiers: Ain Naadja - Baraki will be dug in divided section; cap and stross with a temporary deregister, if it is necessary. However, there are situations of low coverage of land on the vault and/or bad geotechnical conditions where it is necessary to consider the division of the cap in (sides-drifts) In order to maintain the conditions of security and reduce the movements on the surface - settlements.

Between each section, it is necessary to establish excavation shifts so as to reduce the tunnel section which remains excavated without the protection of a suitable retaining to continue to the next phase.

5 The Movements Produced by the Digging of a Tunnel

5.1 Introduction [9]

The construction of tunnels generates disturbances in the internal balance of the surrounding massif, thus creating of deformations and displacement of the soil. It results settlements more or less significant that can affect the stability of neighbouring constructions (buildings and artwork). In urban areas, these disorders can have unacceptable human and economic consequences.

The importance of these settlements is in close relationship with geological and hydrological conditions, the depth of the underground construction and the achieving method.

5.2 General Description of the Soil Movements [4, 5, 8]

The origin of the settlements is complex and, as stated by Leblais (1995) in a recommendation to the working group nb16 of AFTES, the relationship between the settlements generated in surface and in the depth of the work is not easy to determine.

The tunnel digging disrupts the initial field of the constraints and the hydro geological situation in the massif. This modification of the constraints is accompanied in general of an instantaneous moving of the front toward the excavation as well as the walls of the tunnel walls convergence.

The movements of the surface appear in fact as related to a series of concomitant phenomena in the short term.

- Loss of soil in front of size.
- Movement of soil toward the vacuum left, either by the mechanical shovel (or the shield) during its progress, that to say, the distance of installing the support in rear of the front of size.
- Moving of soil to voids
- Deformation of a retaining structure of the tunnel which has just been executed following its loading.

The Fig. 4 presents the movements around the excavation during the digging.

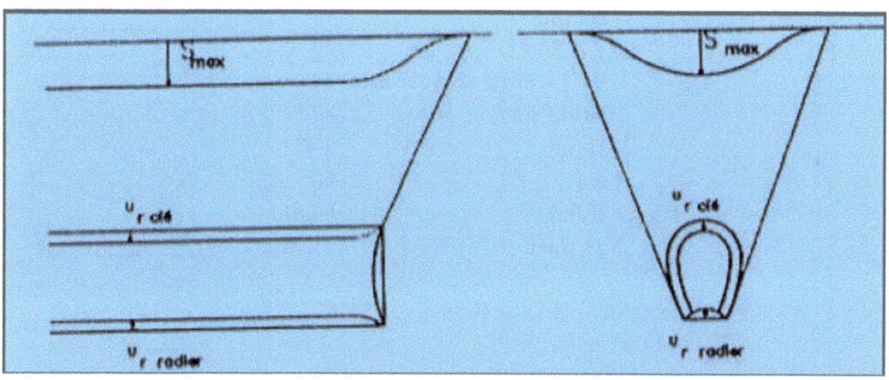

Fig. 4. Displacement of the contours of the excavation according to Leblais (1995)

5.3 Methods of Predicting Settlement

Three approaches have been proposed by Hudson (1976) Allow to estimate and predict the movements in the ground caused by the construction of the tunnel.

- Empirical method: based on numerous detailed measurements (real projects, reduced models).
- Analytical method: This uses analytical formulations based on the equations of the mechanic. These approaches generally require many simplifying assumptions.
- Numerical method: This approach allows calculating the displacement field and constraints in the massif in allowing the simulation of the phases of digging and the taking into account of the laws of behaviour of soil developed and the boundary conditions.

5.4 Digital Application

5.4.1 Calculation of Surface Settlement

The PK 0+11 is located below a villa. For this reason we introduce an additional load P0 generated by the villa.

We use the empirical or semi-empirical method for the calculations of settlement; In general, it aims to estimate mainly the settlements in surface from a low number of parameters, such as the depth of the tunnel, its diameter, the nature of the massif and the loss of volume or the convergence generated by the excavation.

First Method: (Theory of Settlements According to Herzog)

The parameters of land see the Table (5).

Table 5. The parameters of the ground

P0 [kN/m^2]	20
T [m]	12.18
Hü [m]	7.8
γ [kN/m^2]	20
θ [°] = angle de frottement	30
Es [MN/m^2]	42.5
D [m]	9.42
ϕ [°]	60
B [m]	64.87
$f1$ [mm]	21.66

$$\phi = \frac{\pi}{2} - 0$$

$$\mathrm{B} = \frac{D}{COS\,\varphi} + 2 \times \mathrm{T} \times \tan\phi = \frac{9.42}{COS\,60} + 2 \times 12.81 \times \tan 60 = \mathbf{63.22\,m}.$$

$$f_1 = \frac{3\pi}{4}(\mathrm{P_0} + \gamma \times \mathrm{T})\frac{D^2}{B \times \mathrm{E}} = \frac{3\pi}{4}(20 + 20 \times 12.81)\frac{9.42^2}{63.22 \times 42.5 \times 1000}$$
$$= \mathbf{21.48 \times 10^{-3} m}.$$

(Compaction in tunnels according Herzog)
where:

Po: additional load of the surface;
T: distance between axis of the tunnel and surface of the field;
γ: specific weight;
θ: friction angle (φ);
D: Diameter of the tunnel (outside edge of a retaining structure);
B: half width of the settlement trough;
$f1$: height of the settlement trough.

Second Method: Sagasta (1980)

The parameters of land see the Table (6).

Table 6. The parameters of the ground

γ [kN/m^2]	20
D [m]	9.42
N	0.3
E [MN/m^2]	42.5
S_{max}[m]	22.96

$$S_{max} = \frac{\gamma D^2}{E}(0.85 - \nu)$$

$$S_{max} = \frac{20 \times 9.42^2}{42.5}(0.85 - 0.3) = \mathbf{22.96}\,10^3\,\mathbf{m}.$$

where:

E: Module of Young;
ν: coefficient of fish.

- Point of inflection Attwell (1977): because the type of soil is clay,

$$I = 0.5\ H = \mathbf{3.85\,m}.$$

- Basin of compaction (Fig. 5)

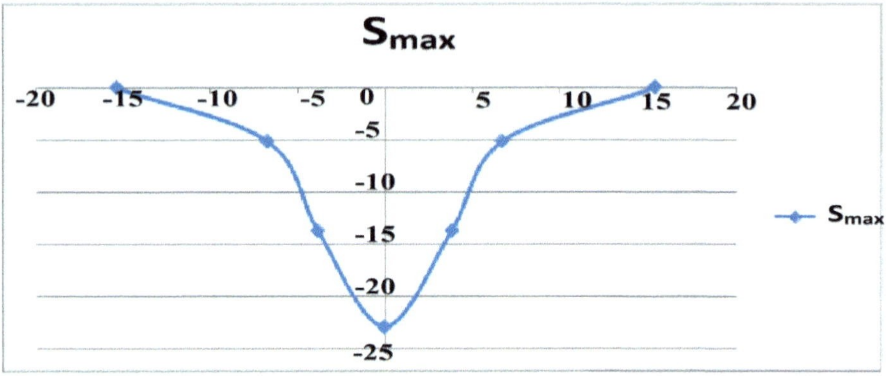

Fig. 5. Cup of compaction

- We use the formula below to see the relationship between the surface settlement trough and the compaction in the key to the Vault:

$$\frac{S_{max}}{S_{clé}} = \frac{\frac{R}{H}}{a + b * \frac{R}{H}}$$

where: R = 4.71 m, H = 7.8 m, a = 0.2 and b = 0.8;

$$\frac{S_{max}}{S_{clé}} = 0.88 \text{ donc } S_{max} = 0.88 S_{clé}.$$

$$\frac{S_{max}}{S_{clé}} = \mathrm{K} * (\frac{R}{H})$$

where:
S_{max}: Maximum settlement;
R, H initial radius and depth of the tunnel.
K is the damping constant:

$$0.1 < R/H < 0.25\ k = 3$$
$$0.33 < R/H < 0.5\ k = 2$$

The hyperbolic formula reflects very correctly the experimental results achieved on rolls of Schneebeli with a = 0.2 and B = 0.8

[C.E.T.U]

- The volume of the cup (according to the formula of Peck (1969)):

$$V_S = 2,5 * i * S_{max}$$

$$V_S = 0.20\,m^3$$

- According to the empirical formula cited by Bernant (1996), we can estimate the relationship between V_S and V_t:

$$V_S = 2,5 * i * S_{max}$$

$$V_S = \frac{\pi}{4} * (D_e^2 - D_t^2)$$

$$\frac{V_S}{V_t} = 0.009 * (\frac{i}{D}) * (\frac{H}{D} - 0.6) = 8.38 * 10^{-4} < 1$$

The report V_S/V_T allows knowing what type of soil we are studying;

- If Vs/Vt < 1 then the soil is dilating;
- If Vs/Vt > 1 then the soil is contractor (Ortigao 1996).

Then the soil is dilating.

6 Overview on the Tool Digital

6.1 Introduction

In this part of work presents the analysis by numerical modeling of the tunnel in several sections.

The main objective of this study is to assess the soil-structure interaction during the digging of the tunnel by the method NATM in a digital simulation of three-dimensional digging of the tunnel.

In order to achieve this result, numerical simulations of three-dimensional were conducted in parallel. The code for the calculation, which was used for this analysis is Plaxis 3D.

6.2 The Code of Calculation Principles [4, 6]

The commercial code Plaxis is based on the finished elements method. This two-dimensional code has been developed first at Delft University of Technology in 1987 to analyze initially the dikes in soft soils. Its field of application is then extended to different types of problems such as the shallow foundations, the support, the embankments and the excavations.

The general algorithm of the Plaxis code is to solve a system of non linear algebraic equations according to an iterative process to determine the fields of shifting to the different nodes in the mesh, the field of constraints and the rupture state of the soil. Once the model implemented on a planar section (Plaxis 2D) or three dimensional (Plaxis 3D) and have defined the lithology of the massif, the code generates the mesh automatically. The elements on which the mesh is based and which have been chosen in the framework of this study are the following (Fig. 6):

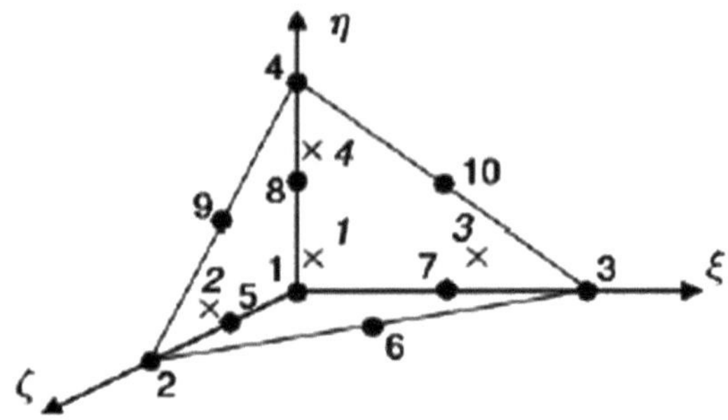

Fig. 6. Basic elements in the mesh size in PLAXIS 3D

- Tetrahedral elements with 10 nodes in Plaxis 3D.

From the boundary conditions and hydraulic and K0 values imposed, the initial constraints state of the massif is calculated. Then the calculation codes allow us to define the different phases which simulate the process of excavation.

Each phase of calculation involves a change in the state of constraints and deformation of the massif. The code then try to achieve a balance for this new state through several steps of calculations, called load "steps". In each step of the calculation, the errors of balance are reduced through a series of iterations. This procedure ends when the new state is reached when there is a break in the soil.

6.3 The Laws of Behavior in Plaxis

There are several models in Plaxis, among the models that we have:

- The model Mohr-Coulomb;
- The model hardening soil.

7 Analysis of Numeric Results and Confrontation with the Measures In-Situ

7.1 Introduction

This work presents the analysis in return through numerical modeling on the measures of the construction site. In the purpose of obtaining a reference model, facilitates the prediction of the surface settlement in other mileage points having the same characteristics. Modeling is done by the Plaxis three-dimensional code.

7.2 Soils Properties

The Table (7) summarizes the geotechnical characteristics of different layers of soil to the PK 0+220 of tunnel in the segment D9.

Table 7. Characteristics of soils

	Embankment (RX)	Marne yellow (QM)	Yellow sand to brown (TS)
Model/Type	Mohr –Coulom drained	Mohr-Coulomb not drained	Mohr-Coulomb drained
Volumetric weight			
γ_{unsat}	22.8	24	25.2
S_{aut}	22.8	24	25.2
Permeability			
K_x	0	2 E^{-8}	1 E^{-4}
Ky	0	2 E^{-8}	1 E^{-4}
Parameters			
Module of the deformation $E_{ref} = E'$	6 E^3	3 E^4	8 E^4
Module of fish $\nu = \nu'$	0.2	0.2	0.2
Cohesion $C_{ref} = c'$	5	150	15
Friction angle $\varphi = \varphi'$	28	27.5	33
Coefficient of the expansion $\psi = \psi'$	0	0	0.3
The interfaces			
Rigidity R_{inter}	0.67	0.67	0.67
Coefficient of pushes of land at the repose			
$k0 = 1 - \sin\phi$	0.53	0.63	0.46

7.3 Establishment of a Reference Model for the Subway of Algiers

This part this numerical modeling (3D) of mileage point [0+11, 0+24] which is located in the Section D9 (Ain Naaja -Baraki). In order to estimate the soil- structure interaction and the ground movements generated by the digging of tunnel in an urban environment.

In order that the Plaxis software can do the calculations correctly and completely, we must enter all the data in the following project:

7.3.1 Three-Dimensional Model

Part I: Input

Geometric and geotechnical hypothesis

- **Geological and geometric cut and of model**

The geological cut shows that we have three layers of soil component the geology of the site (Fig. 7):

- A layer of embankment in surface of 6.6 m of depth;
- A layer of yellow marine from El Harrach 10 m of thickness;
- Finally a layer of yellow sand with passage of sandstone of 13.4 m of thickness.

- **The geometry of the model**

The project will be modeled by a three-dimensional geometric model 50 m wide(X) on 30 m depth (Y) with an extension (Z) of 50 m so that we can neglect the influence of edges (Fig. 8).

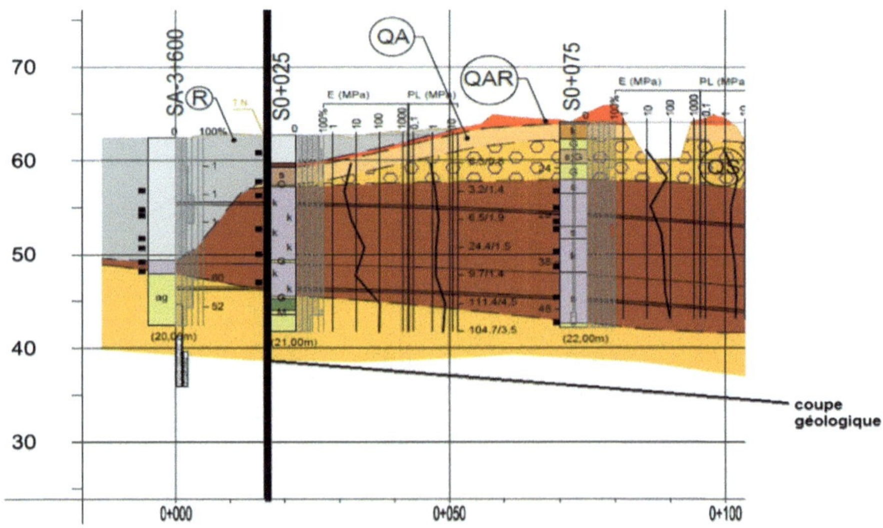

Fig. 7. The geological profile of Section D9

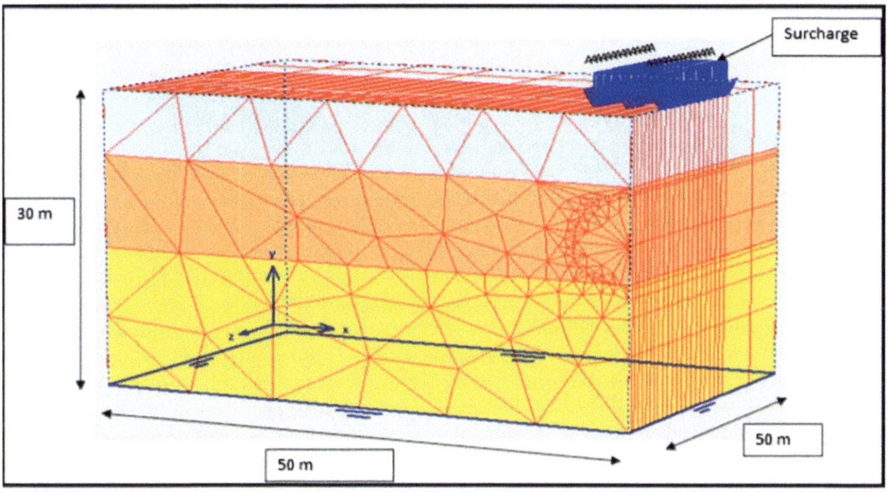

Fig. 8. The geometry of the model in 3D

- **3D mesh generation**

The digging begins from mileage point [0+12, 0+20], the phase modeling of dig has been designed in accordance with the steps for digging really made, where the digging has been achieved by divided sections with a gap of (2 to 5 m).

That is to say we excavated 8 m; the gap between the cap and the Stross/foundation is 2 to 5 m (Fig. 9).

Fig. 9. Diagram showing the advancement of digging

The coverage above the tunnel always remains 7.8 m. The mesh is refined around the excavation (tunnel).

In the end, the field is modeled by prismatic triangular elements of 15 nodes with approximately 4893 elements, and 13879 nodes (Fig. 10).

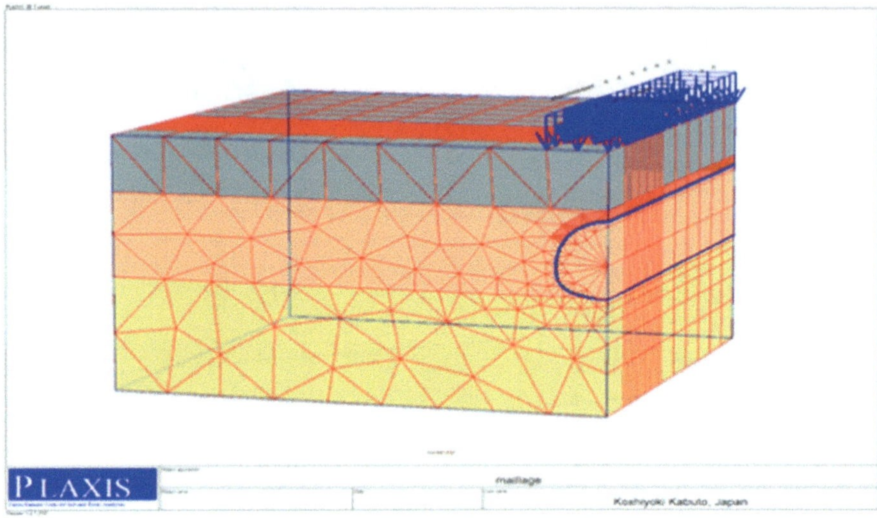

Fig. 10. 3D model adopted, 3D mesh view

- **Generation of the pore pressure**

In this project no groundwater is present, so we put the groundwater table at the level zero of the model (Fig. 11).

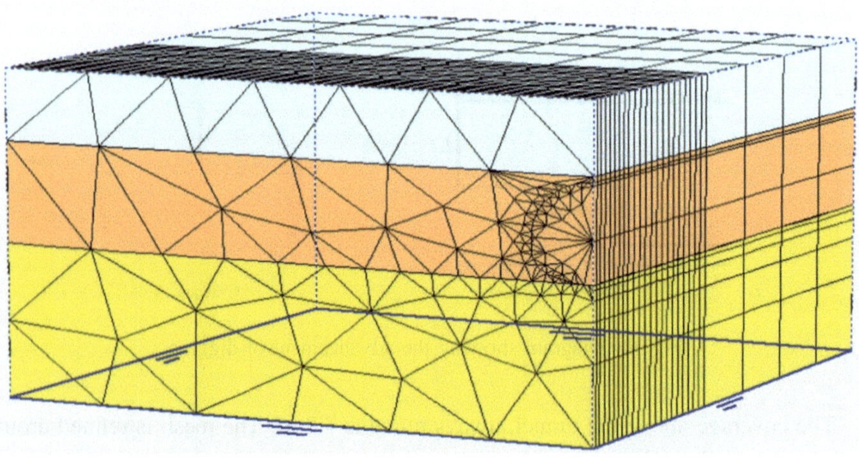

Fig. 11. Interstitial pressure condition

Part II: Calculation

Modeling of the tunnel excavation NATM

- The different rates of containment in the phases of simulation according to the Table (8).

Table 8. Deconfine rate values

Advancement (X in m)	Tunnel radius (R in m)	Deconfine rate (β)	ΣM placement (insert in Plaxis 3D)
cap (1) 0.8 m	5.01	0.49	0.51
cap (1) 1.6 m	5.01	0.63	0.37
cap (1) 2.4 m	5.01	0.72	0.28
cap (1) 3.2 m	5.01	0.78	0.22
cap (1) 4 m	5.01	0.82	0.18
cap (1) 4.8 m	5.01	0.85	0.15
cap (1) 5.6 m	5.01	0.87	0.13
cap (1) 6.4 m	5.01	0.89	0.11
cap (1) 7.2 m	5.01	0.91	0.09
cap (1) 8 m	5.01	0.92	0.08
Stross/radier 0.8 m	3.5	0.41	0.59
Stross/radier 1.6 m	3.5	0.61	0.39
Stross/radier 2.4 m	3.5	0.79	0.21
Stross/radier 3.2 m	3.5	0.84	0.16
Stross/radier 4 m	3.5	0.88	0.12
Stross/radier 4.8 m	3.5	0.9	0.1
Stross/radier 5.8 m	3.5	0.93	0.07

- **The phases of calculation**

Phase 0: Initial phase;

Phase 1: The model remains in the initial state;

Phase 2: activation of the load (25 kPa) in the section where the Villa(R+1) is situated;

Phase 3-phase 17: The cap excavation, and the activation of a retaining structure of the cap with an advancement of 0.8 m in each phase;

18 phase-phases 25: Stross/foundation excavation and the activation of a retaining structure of Stross/foundation with an advancement of 0.8 m in each phase.

At the end of these previous phases we obtain the following model (Fig. 12).

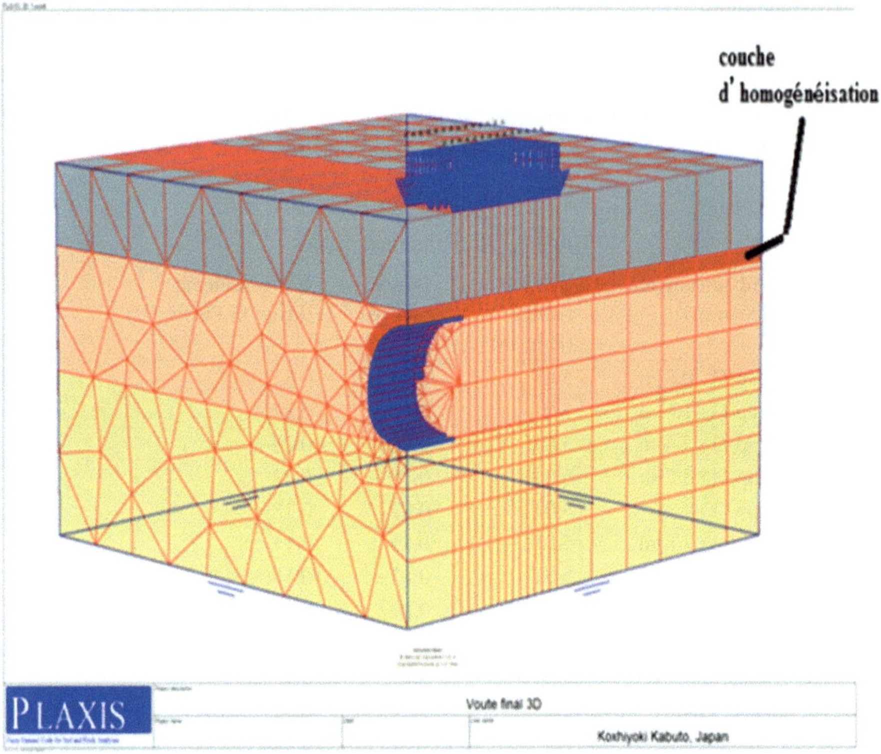

Fig. 12. The shape of the 3D model of the project during the excavation

Part III: Results

In this part we will present the results obtained:

- Figure 13 represents the settlement made by building at the initial state;
- We note that there is an interaction building/tunnel;
- The most important investments appear below the overload of the building and minimize in approaching the vault of the tunnel;
- We note also an important shift in the front line of the tunnel.

Interpretation
The vertical displacement is at maximum below the building and decrease in function of the depth.

The vertical displacement impacts on neighboring soil of the tunnel and this influence increases slightly at the level of the walls of the tunnel.

Note
There is a temporary foundation between the cap and the Stross in order to limit the vertical displacements during excavation (Figs. 14 and 15).

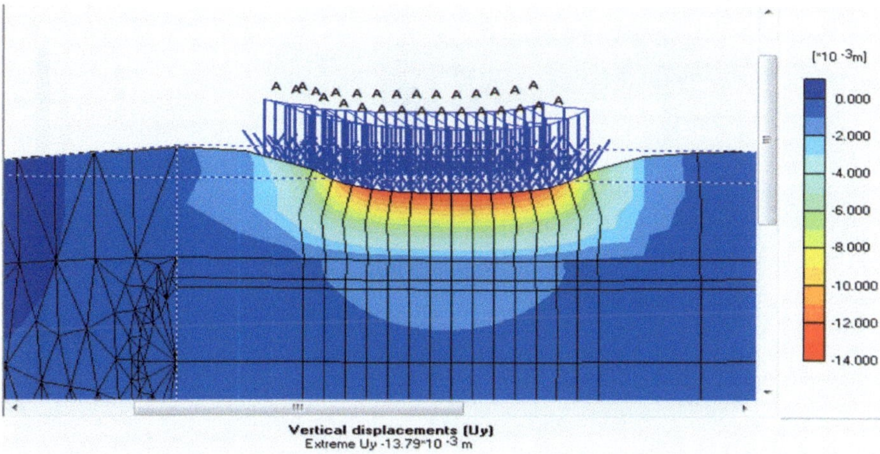

Fig. 13. The settlement carried out under building in the initial state

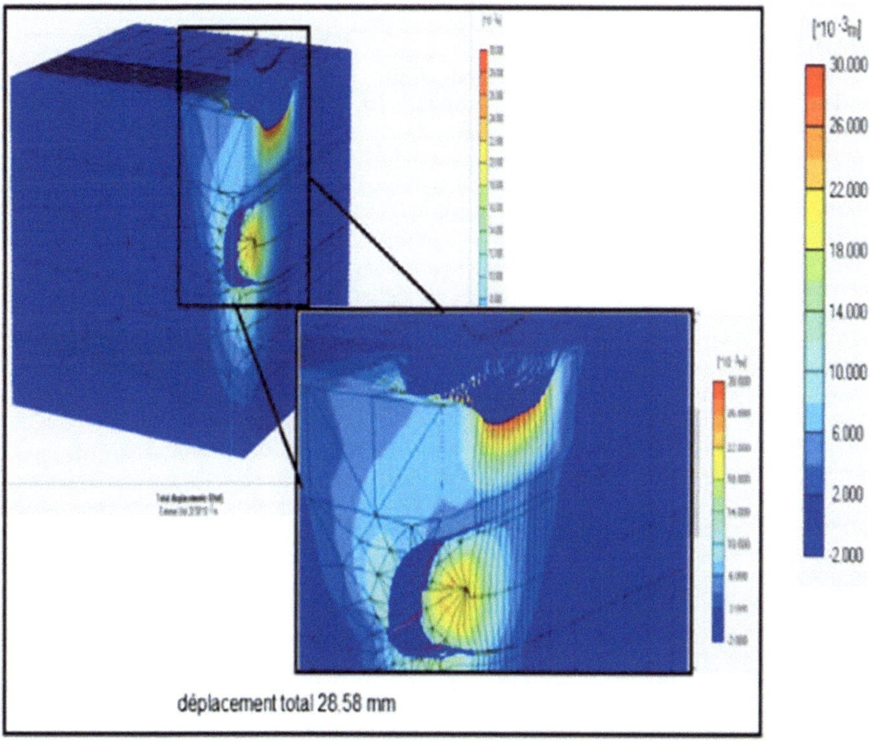

Fig. 14. Total displacement of the structure

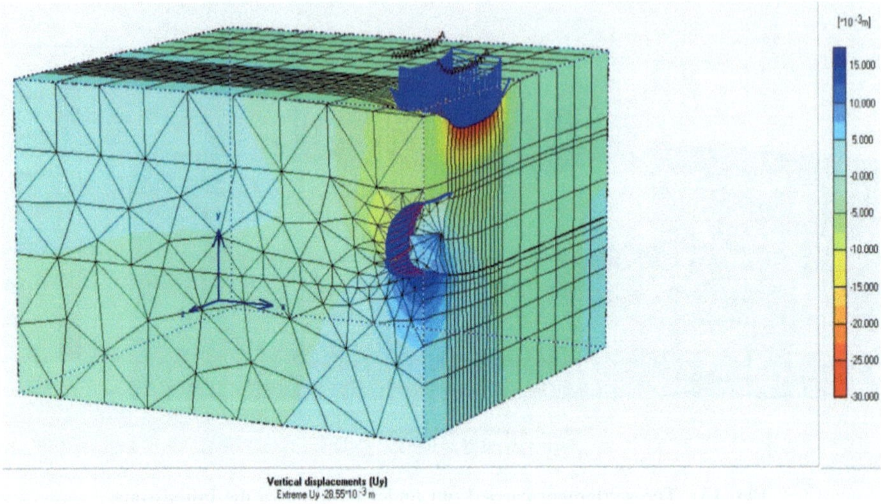

Fig. 15. Representation of the vertical movement

7.3.2 The Movement

1-As Keystone: In the Fig. 16.

The total displacement is: $U_{TOT} = 8.75$ mm;
The next move the x axis is: $Ux = 1.87 \times 10^{-3}$ mm;
The next move y axis: $Uy = 8.72$ mm.

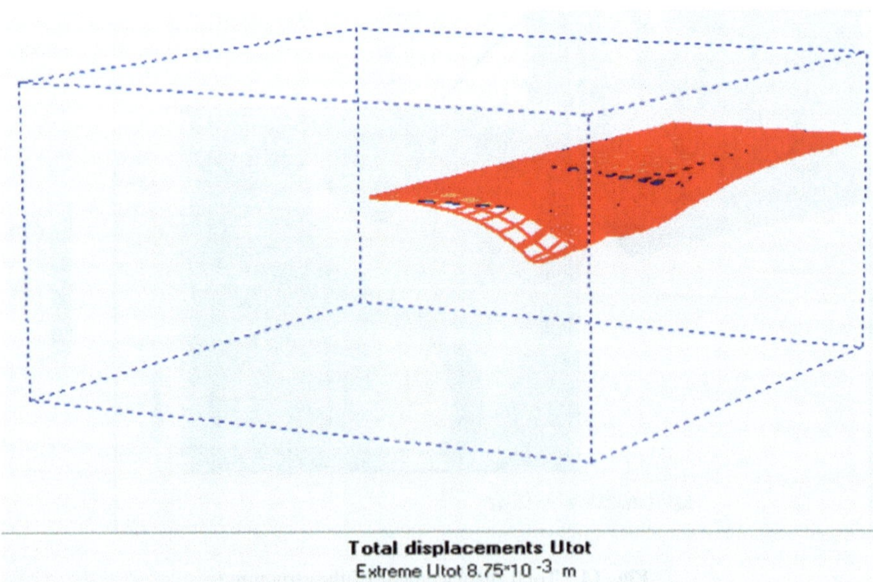

Fig. 16. Bowl at the level of the to the vault

According to the results obtained, we note that the vertical movement is moderately low at the level of the tunnel vault.

2-Surface: In Fig. 17 we present the cup of surface settlement.

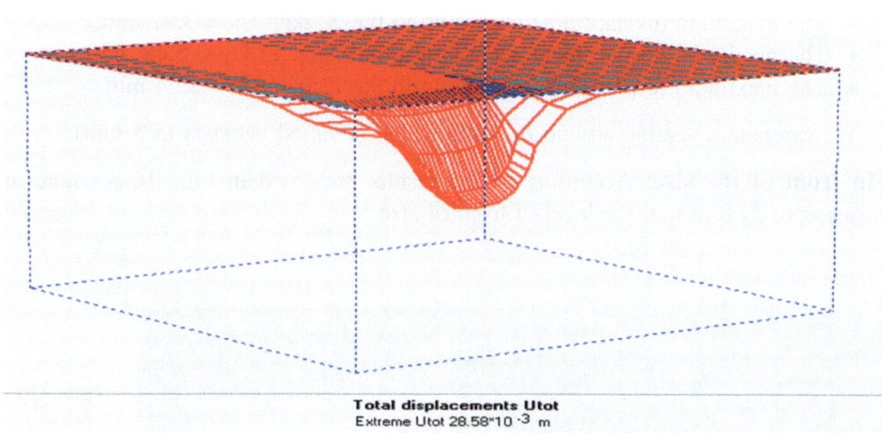

Fig. 17. Representation of the settlement cup

- The maximum vertical displacement curve of the building.

We select the point to which the settlement is maximum that is to say below the building (Fig. 18).

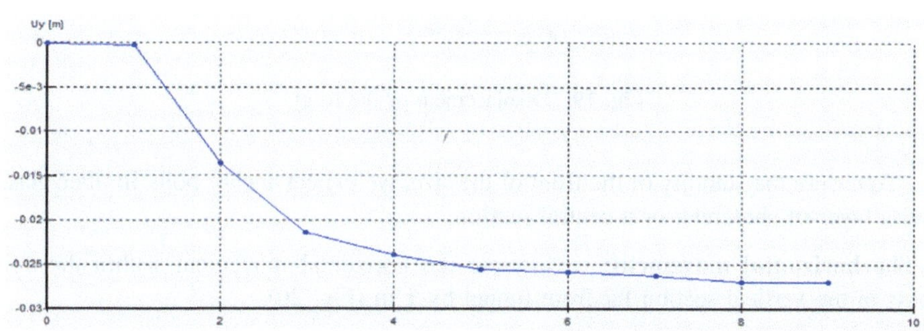

Fig. 18. The maximum settlement of the building during the digging

a. The part where there is the overload of villa (25 KPA):

- The total displacement: 28.58 mm;
- The maximum displacement according to the X axis: Ux = 4.39 mm;
- The maximum displacement according to the Axis Y: Uy = 28.58 mm;
- The maximum displacement according to the Z axis: Uz = 5.57 mm.

We can say that the settlement (vertical displacement) varied between 17.26–28.58 mm in the area where there is the load of villa.

b. The neighboring part (takes 1 m) of the villa:

- The total displacement: 6.56 mm;
- The maximum displacement according to the X axis: Ux = 3.51 mm;
- The maximum displacement according to the Axis Y: Uy = 5.48 mm;
- The maximum displacement according to the Z axis: Uz = 2.24 mm.

We can note a settling around the villa which is varied between (5-8 mm).

3-In front of the size: According to the results presented in Fig. 19, we note an extrusion of 21.8 mm at the level of front of size.

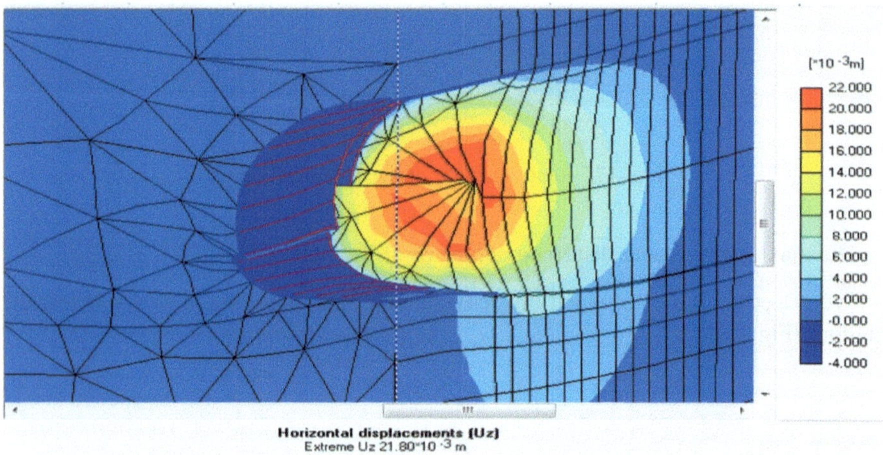

Fig. 19. Displacement of the front

To ensure the stability of the front of the size, we expect anchor bolts in fiber glass and a layer of shotcrete, or a central merlon.

4-The horizontal movement: we present the horizontal movement within the soil mass in the vertical section far from tunnel by 1 m (Fig. 20).

- The plastic points (Fig. 21).

Note
The plastic points determining the critical plastic state.

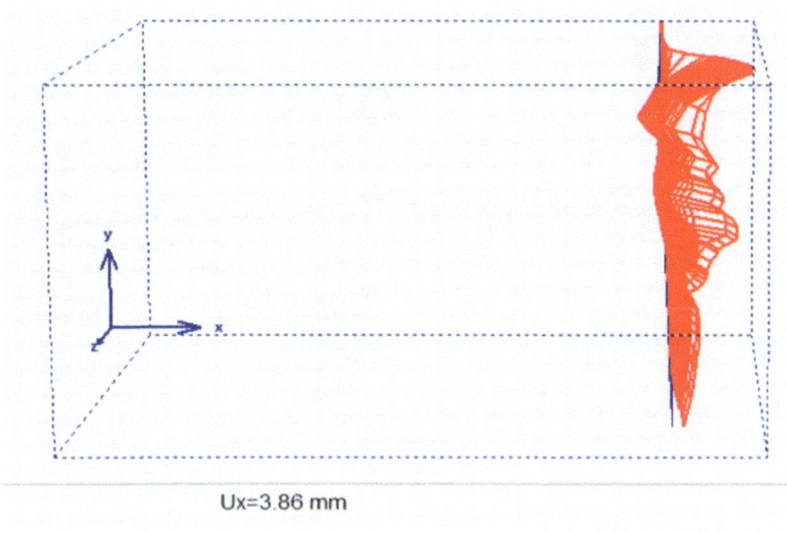

Fig. 20. The horizontal displacement along a vertical section (3.86 mm)

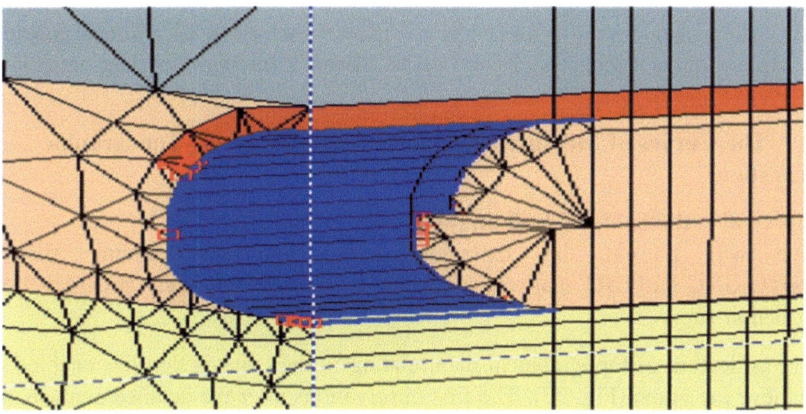

Fig. 21. Representation of plastic points

7.4 Settlement Control by the Targets Monitoring in the Construction Site

It is necessary to place monitoring instruments to follow up the compaction during digging a tunnel in urban areas, as shown in the Fig. 22.

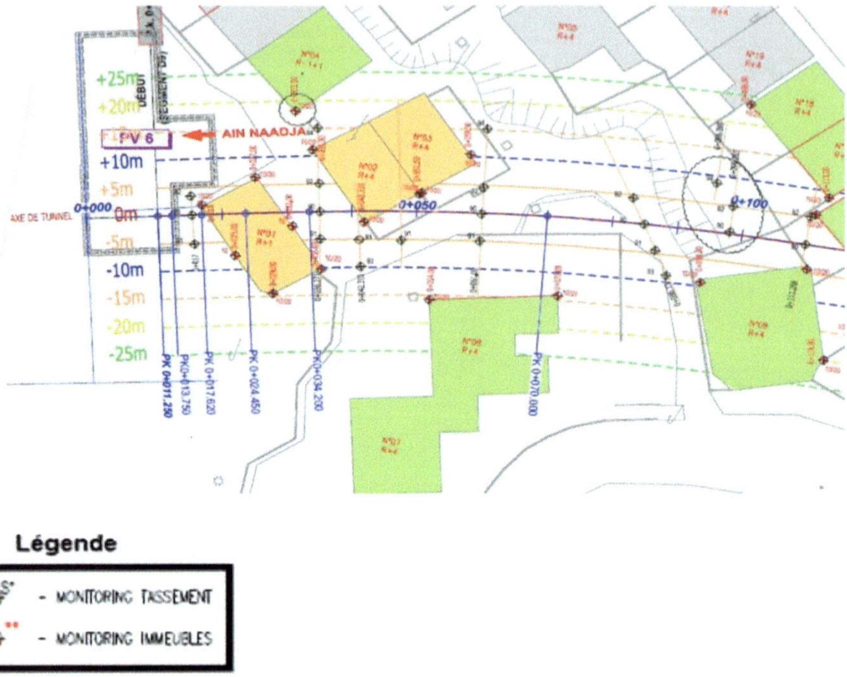

Fig. 22. The positions of the monitoring targets (Note * Each target has a specific coordinated. * The position of the auscultation devices can be adapted according to existing constraints.)

7.4.1 The Curves of Measured Settlements In-Situ by the Topographic Observations

1-Curves of settlements of Building 01:

We will present the settlement curves of the building 01 from MONITORING REPORT that was conducted by the office of study at the level of the construction site.

- The collection of topographical data relating to the construction has enabled us to achieve the graph (Fig. 23). The first observations after the detailed analysis of the results are the following:

  The curve indicates that well before the beginning of excavation of the segment D9, a previous settlement was done. This can only be explained by the achieved preparatory close to the building and the influence that they are likely to have on the ground.

  The graph indicates that the settlement values show a significant and proportional increase of the work progress. However, the changes pose a cyclical character, intersected with a period more or less stable which is the result of the break of a retaining structure. The latter tends to stabilize toward a final value equivalent to 30 mm, which corresponds to a stop of the digging work followed by strengthening the ground.

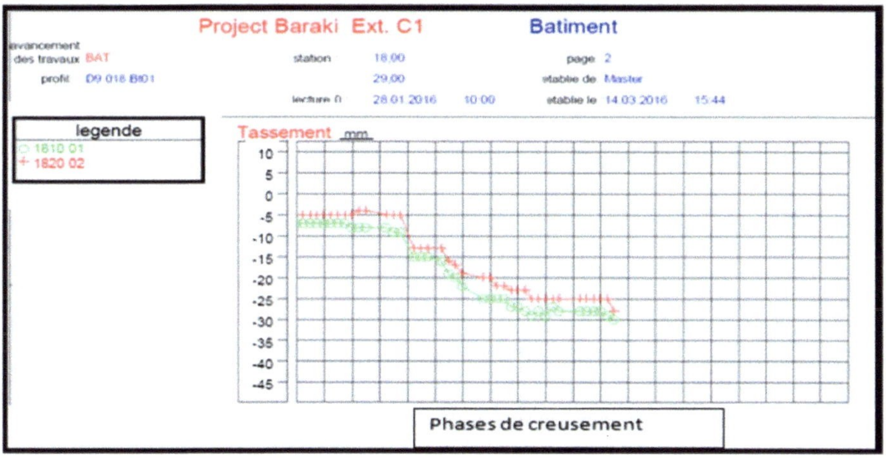

Fig. 23. Building 01settlement

2-The settlements curves of surface surrounding the Building 01:

The Fig. 24 presents a low surface settlement around the building 01 which does not exceed 5 mm.

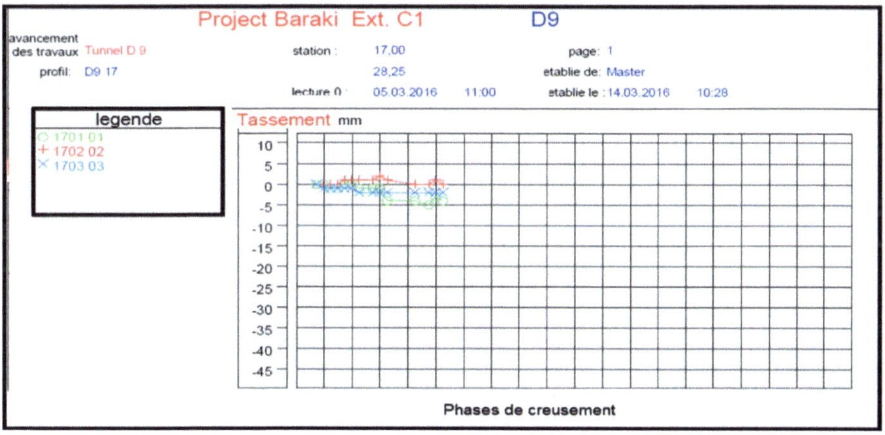

Fig. 24. Surface settlement

7.4.2 Comparison Between Different Results

We are going to do a comparison between the settlement results achieved by measures in-situ and others obtained by numerical calculations in the purpose of reference model.

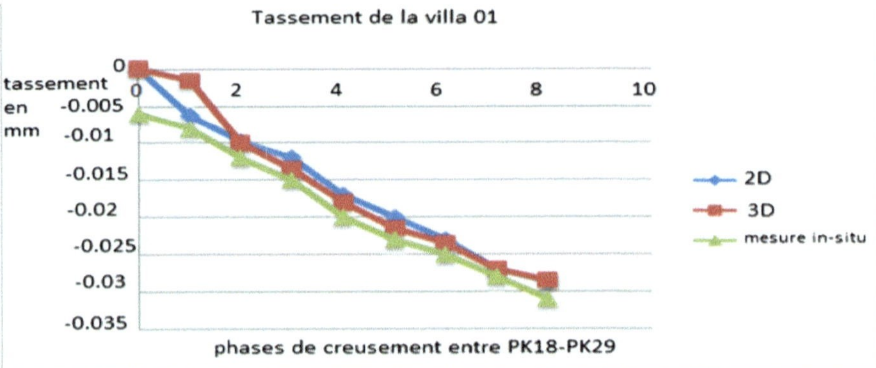

Fig. 25. Comparison between the results of model 2D-3D and measures in-situ

- The curve of Fig. 25, highlights the different results of the numerical simulation and the measures in-situ.

We can see the strong resemblance between the numerical predictions and the values in-situ. Despite the apparent coincidence of the models 3D and 2D, the three-dimensional simulation presents a more specific approach of the reality, it attaches more importance to the different constraints and deformations.

7.5 Damages Calculation

In the checking of the damage caused to buildings, they are considered as structures sufficiently flexible to accompany the subsidence profile, therefore the parameters characterizing the risk of damage to structures are estimated in this situation.

The most important definitions of this analysis, based on the displacement (settlements) of four points A, B, C, and D under the foundations (isolated or continued) of a building, are presented in the Fig. 26.

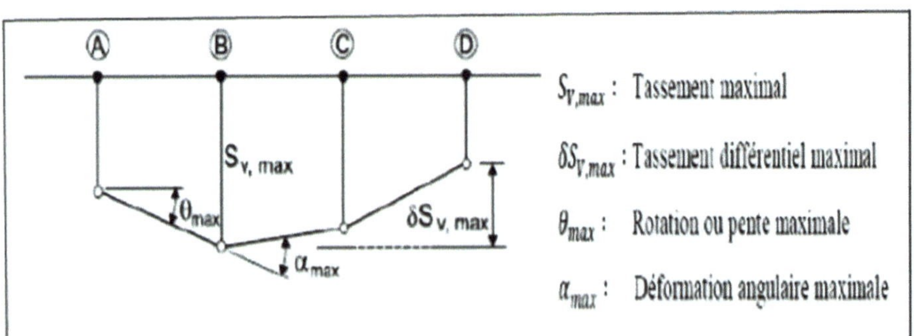

Fig. 26. The different settlement points in the load area

7.6 The Modeling Results

- **Displacement**

1. The total displacement: See the Fig. 27.
   The total displacement is about 44 mm.

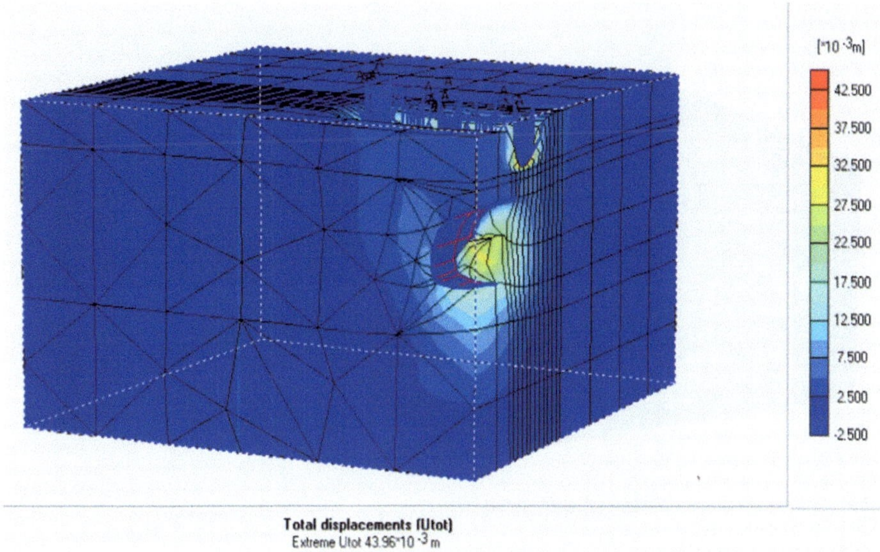

Fig. 27. Total displacement

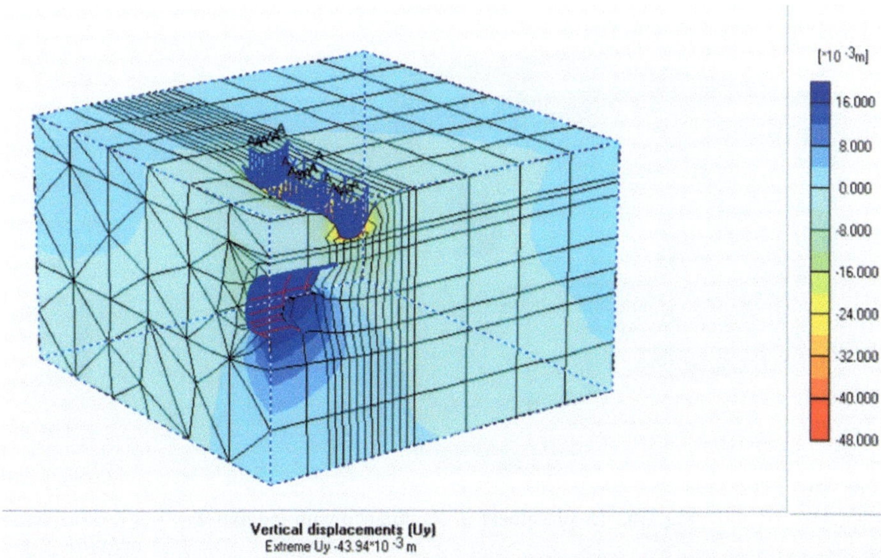

Fig. 28. Vertical displacement

2. Vertical displacement: See the Fig. 28.
   The vertical displacement is 44 mm.
3. Horizontal displacement following the X axis: See the Fig. 29.
   The next displacement of the axis X = 11.16 mm.
4. Horizontal displacement following the Z axis: See the Fig. 30.

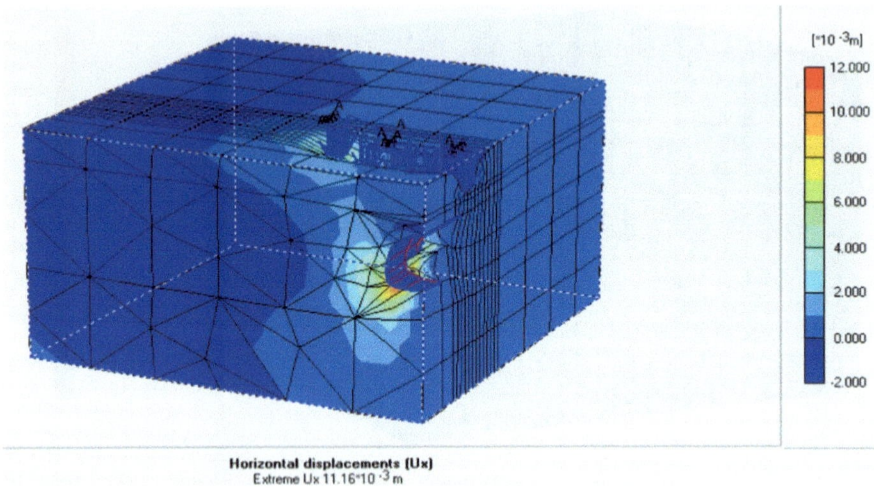

Fig. 29. Displacement in accordance with the X axis

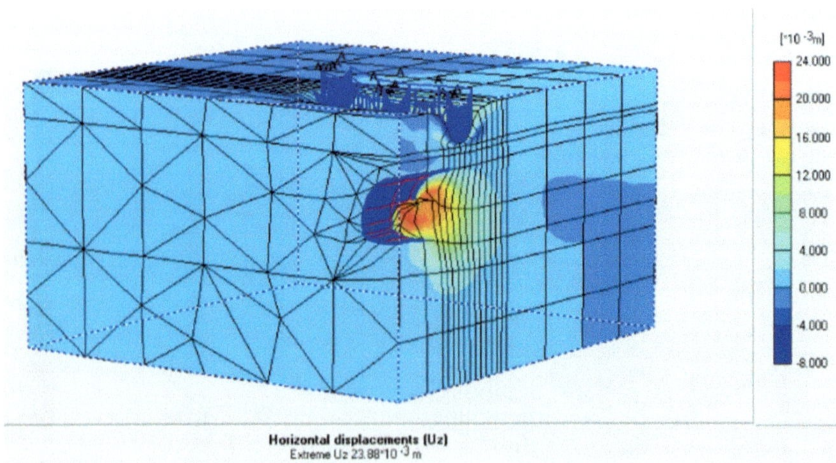

Fig. 30. Displacement in accordance with the axis Z.

8 Comments

We note that the vertical displacement is at maximum below the building, this importance of displacement where compaction is linked to the phasing of digging, and the weakness of the ground. Therefore, if necessary to strengthen the overlying ground with the building before the beginning of excavation.

9 Conclusion

The conflicts between the results of auscultation and the numerical simulation show that the use of 3D model allows obtaining results in agreement with the real phenomena observed during the digging of the tunnel.

The results of surface settlement are relatively high. Despite the fact that it does not represent any threat to the underground building as the reliability of the support (Umbrella vault), it is very worrying because it puts in danger the building located above. It is therefore logical that we draw a few recommendations to strengthen the land in place and prevent any further aggravation of the situation. Here are some:

- Strengthening the ground with columns of jet-grouting.
- Injections of cement grout in the soil between the tunnel vault and the footings foundation of the building.
- Construction of a general slab strike under the building.

10 Special Dedication

To the candle who has always been burned to illuminate my paths, to the sun from which I inspire my life. You, Mom, I dedicate you my accomplishment, May Allah prolong your life. To the lamp, who always been the light in the darkness of my nights, to you brother Abdennoor, I dedicate my accomplishment. To whom I have grown up on her nestle, my sister Malika and her husband Nasser. To my sisters who have been the energy that has given me the power to achieve success, Massaouda, Wardishen, my little Sabrina. To whom I fully share everything and he is everything for me, my brother Hamza. To my brothers: Mohammed and Assyl. To my nephews: Wahadj, Mouhib, Moughith, my pearl Jannah, my sapphire Mouhja. I dedicate this work to my uncle Salah Chouder, school of life, to my uncle Messaoud Amroune. To all my friends, but have no enough space to mention them all.

Acknowledgement. First of all, praise to Allah who, thanks to His help and His will, has brought me to the end of this work by preserving me a healthy life during all my studies. I would like to sincerely thank all those who me to carry out this work.

Thanks To my doctoral supervisor, Mr. HAFSAOUI Abdellah for his great availability, moral and intellectual support during the realization of this work.

References

1. Bouvard-Lecoanet, A., Esteulle, F., Colombet, G.: Ann. Underground Works (1998)
2. Center of Studies of Tunnels: The design and sizing (Section 3), Paris (1998)
3. EMA: Note of calculation. Tunnel in mine, temporary support, Section D8.1, Algiers (2012)
4. Fethi, K.: Numerical study of the stability of a tunnel. Section T4 of the East-West Highway, Algiers (2012)
5. The analysis and forecasting of surface settlement during the digging of the north tunnel of the underground crossing of Toulon, Toulon (2002)
6. Numerical modeling of a tunnel digging with the assistance of a mud ground pressure. The National School of the Roads and Bridges, Paris (2002)
7. Amar, S., Jezequel, J.-F.: Mechanical properties of determined soils. Techniques of the Engineer, Paris (1998)
8. Tunnels in urban areas. Forecasts of the settlements taking into account the effects of the Pre – supports (strengthening the front of size and vault- umbrella). The National Institute of Applied Sciences of Lyon, Lyon (2012)
9. Hafsaoui, A.: Design course of underground works. Badji Mokhtar University, Annaba (2016)

Mitigating Foundation Settlement Induced by Tunnel Construction

Mona Badr El-Din Anwar(✉)

Civil Engineering Department, The German University in Cairo, New Cairo, Egypt
mona.anwar@guc.edu.eg

Abstract. In urban regions tunnel construction has become common to enhance transportation networks and crossing crowded areas or to construct underground metro lines. Tunnel construction in urban areas usually passes in the vicinities of building foundations. Unless sufficient clearances exist, shallow tunneling can cause settlements and/or tilting of adjacent foundations. In most cases, this clearance is not affordable. One of the proposed solutions to reduce the effect of tunnel construction on existing foundations is to use barrier walls to reduce soil movement under the foundations. In this research, 3D finite element analyses are carried out to investigate the influence of using different types of barrier walls in mitigating foundation settlements due to tunnel construction. Pile walls with different diameters, spacing, and lengths are studied. Diaphragm walls with different thicknesses and lengths are also investigated. The resulting settlements of the foundations are presented and discussed.

Keywords: Tunneling · Barrier wall · Settlement · Tilting

1 Introduction

Tunnel construction is increasing each day to enhance transportation networks. In urban areas, tunnel alignments become a real challenge due to its proximity to existing structures. In some cases, the horizontal spacing between the tunnel center and the foundation is less than the tunnel diameter (Cheng et al. 2004). Tunnel construction is always accompanied with non-uniform ground settlement which threatens the safety and stability of those nearby buildings and infrastructures (Gan et al. 2014). Hence, it is necessary to evaluate the ground movement caused by tunneling (Jin and Yang 2014; Aversa et al. 2013; El-Nahas et al. 1997). Yet, accurate prediction of tunneling effects on nearby structures and services in the urban environment poses a major challenge during design (Lee and Ng 2005). Many analytical solutions have been given to predict the ground movement due to tunneling works in both clays and sands. In practice, the most commonly used method is the Gaussian curve suggested by Peck (1969). It is worth noting, and as revolved from model test results and field tests, that the obtained settlement trough in sand is different than that obtained in clay (Jin and Yang 2014).

The original version of this chapter was revised: Author's second affiliation has been deleted. The erratum to this chapter is available at 10.1007/978-3-319-61636-0_17

S. Agaiby and P. Grasso (eds.), *Engineering Challenges for Sustainable Underground Use*, Sustainable Civil Infrastructures, DOI 10.1007/978-3-319-61636-0_15

The effect of tunneling works on a nearby deep foundation has been studied by many researchers, (e.g. Gerheim et al. 2015; Cheng et al. 2004, Dias and Bezuijen 2014; Hu et al. 2013). Despite that those researches were interested mainly in the behavior of deep foundations, yet they highlighted that the soil behavior during tunneling works was affected by the presence of piles. Furthermore, they found that the change in piles stiffness, length, horizontal spacing from tunnel and piles spacing affected the overall behavior of the soil around the tunnel and affected the shape and value of the settlement trough. The presence of piles alters the greenfield deformation patterns around the tunnel. While in greenfield conditions, the soil at the depth of the tunnel is attracted toward the cavity, but the piles, which are founded below the tunnel invert, reduce such a pattern of movement (Bilotta and Russo 2011).

In other cases, piles have been sometimes purposely installed before tunneling to protect existing foundations from excessive movements. Di Mariano et al. (2007) reported case study on using 650 mm bored piles wall, with spacing between piles equals to twice the pile diameter (i.e. $2D_{pile}$) and piles tip level were equal to the tunnel invert level, in front of existing building as a measure of protection from ground deformation during tunnel construction. It was found that the wall provided a barrier to movement between the tunnel and the building and reduced the soil volume loss. In 2015, Alonso and Ledesma presented an overview of the railway tunnel crossing the city of Barcelona. The tunnel is in proximity of a monumental temple façade. To protect the temple façade from any expected settlement, a barrier wall was suggested to be built in between the tunnel and the existing deep foundation which was supporting the façade. The suggested wall consisted of piles 1.5 m diameter and 2 m spacing (center to center). Piles length was 41 m, which extended below tunnel invert level. It was found that the measured settlement trough was much less than the predicted at the design stage.

In this study, in attempt to investigate the effect of installing barrier wall between tunnel and adjacent structure, a finite element analysis using PLAXIS 3D 2016 is carried out. The finite element model investigated the difference between using disconnected piles, connected piles and diaphragm wall, as a barrier, in mitigating the structure settlement due to tunnel construction. Different piles diameter, spacing and length were studied for both disconnected and connected piles. Different wall thickness and length for the diaphragm wall were also studied. Results were analyzed and discussed to present the obtained behavior.

2 Soil and 3D PLAXIS Model

Three-dimensional analysis was carried out using PLAXIS 3D 2016 software. Hardening soil model was considered to represent the soil behavior. Hardening soil model is an advanced soil model. It is a non-linear elasto-plastic model that accounts for stress dependency of the soil stiffness in both loading and unloading conditions Schanz (1998). Tunnel was modeled as circular tunnel with outer diameter D_{tunnel} of 8 m and tunnel length was 15 m. The TBM advancement was modeled using initial lining with thickness 0.25 m. This tunnel diameter is considered the average value for typical underground railways (Bilotta and Russo 2011; El-Nahas et al. 1997). Tunnel center is 12 m below ground surface which gives cover to diameter ratio equal to unity

($C/D_{tunnel} = 1$). Volume loss of 2% was adopted as it can be considered average value for volume loss in reasonably good tunnel construction technique and workmanship (Cheng et al. 2004; Minh et al. 2016; Di Mariano et al. 2007). The volume loss was modeled by the contraction stage during tunnel construction. The grouting pressure was taken equal to the overburden pressure to keep soil body from collapsing inside the tunnel and finally the tunnel body was represented by concrete elements.

Three types of barrier walls were investigated between the tunnel body and the adjacent building, separated piles, connected piles and a diaphragm wall. Embedded beam elements were used to model the piles in the first and the second types of wall. Embedded beam is a structural object composed of beam elements that can be placed in arbitrary direction in the sub soil that interacts with the sub-soil by means of interface elements. Pile diameters of 0.5 and 0.6 m were considered. Piles spacing were $2D_{pile}$, $4D_{pile}$ and $6D_{pile}$, and pile wall horizontal distance from tunnel center was considered as $1D_{tunnel}$. Different pile lengths were considered to investigate the optimum extend of pile wall. For this study where ($C/D_{tunnel} = 1$), the pile lengths were considered as $1D_{tunnel}$ (i.e. 8 m piles length, ended above tunnel top level), $1.5D_{tunnel}$ (i.e. 12 m piles length, ended at the centerline level of tunnel) and $2D_{tunnel}$ (i.e. 16 m piles, ended at the tunnel bottom level). The second type of wall was similar to the first but with a top beam connecting the piles to investigate the effect of pile connection on the wall behavior.

Solid element with interface elements was used to model the diaphragm wall. Three thicknesses were considered for the wall (0.3, 0.5 and 0.6 m). Depth of wall varied from $1D_{tunnel}$, $1.5D_{tunnel}$ to $2D_{tunnel}$.

Table 1. Material parameters for soil and final tunnel body

Parameter	Symbol	Sand	Concrete	Units
Material model	Model	Hardening soil	Linear elastic/non porous	-
Unit weight	γ	17	24.7	kN/m^3
Secant stiffness	E_{50}^{ref}	25×10^3	2×10^7	kN/m^2
Tangent oedometer stiffness	E_{oed}^{ref}	22×10^3	-	kN/m^2
Unloading/reloading stiffness	E_{ur}^{ref}	75×10^3	-	kN/m^2
Cohesion	c'	0	-	kN/m^2
Friction angle	ϕ'	30	-	Deg.
Dilatancy angle	ψ	0	-	Deg.
Poisson's ratio	v	0.3	0.12	-
Interface reduction factor	R_{inter}	0.7	-	-
Lateral earth pressure	k_o	0.5	-	-

Table 2. Material properties of the plate representing the TBM and building foundation

Parameter	Symbol	TBM	Foundation	Units
Thickness	d	0.25	0.5	m
Unit weight	γ	24.7	24.7	kN/m^3
Material behavior	-	Linear elastic	Linear elastic	-
Young's modulus	E	2×10^7	2×10^7	kN/m^2
Poisson's ratio	v	0.12	0.12	0.12

Plate element was used to model a building foundation at ground level, in the vicinity of the tunnel. Surface load was added to represent a building with bearing stress 100 kPa. Building stiffness was not considered in this study to simplify the model and reduce the time of analysis. Tables 1, 2 and 3 present the adopted parameters and Fig. 1, presents the different elements in the model.

Table 3. Material properties of piles and the connecting beam

Parameter	Symbol	Pile	Units
Young`s modulus	E	2×10^7	kN/m^2
Unit weigth	γ	6	kN/m^3
Pile type	-	Predefined	-
Predefined pile type	–	Massive circular pile	-
Diameter	D_{pile}	0.5 and 0.6	m
Skin resistance	Type	By layer	-
Maximum friction	$T_{skin,\ max}$	200	kN/m`
Base resistance	F_{max}	300	kN

Parameter	Symbol	Beam	Units
Young`s modulus	E	2×10^7	kN/m^2
Unit weigth	γ	6	kN/m^3
Area	-	0.4 – 0.48	-
Material behavior	–	Linear	-
I_3		0.042	m^4
I_2		0.01 – 0.018	m^4

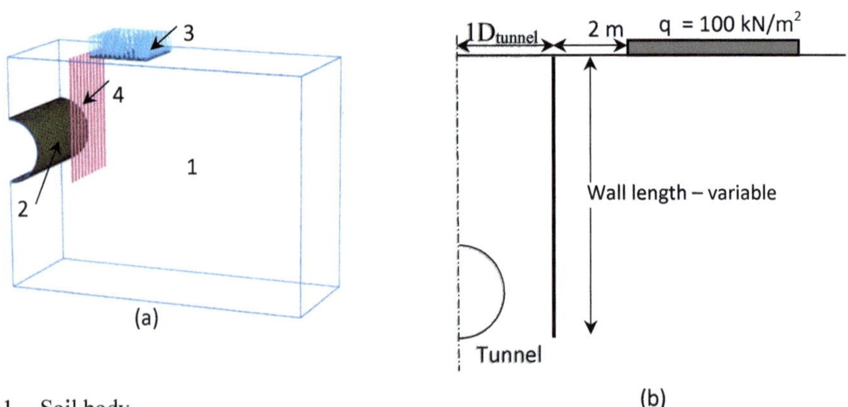

1. Soil body
2. Tunnel body (D_{tunnel}=8m)
3. Loaded Plate (8 X 8m) representing small raft on ground surface at distance = 1.25 D_{tunnel}
4. Wall installed at spacing 1D_{tunnel} (i.e. 8m) from tunnel center :
   a. Wall type 1: Piles wall. pile spacings change S/D_{pile} = 2, 4 and 6 and pile lengths are 1D_{tunnel}, 1.5 D_{tunnel} and 2D_{tunnel} (i.e. 8, 12 and 16m)
   b. Type 2: Piles forming a wall and piles are connected with head beam
   c. Type 3: Concrete diaphragm wall. Wall thicknesses are 0.3, 0.5 and 0.6m. Wall

Fig. 1. Model elements

3 Ground Settlement Trough

Peck 1969 presented the most common prediction for surface settlement trough in tunnel design, which is the normal probability distribution curve. In this curve, the maximum settlement occurs above the tunnel centerline, as illustrated in Fig. 2. As the tunneling is advancing, a settlement trough develops. In case of structure exists within the predominance of the developed settlement trough, this structure will be affected, and the presence of this structure will consequently modifies this trough (El-Sayed 2001).

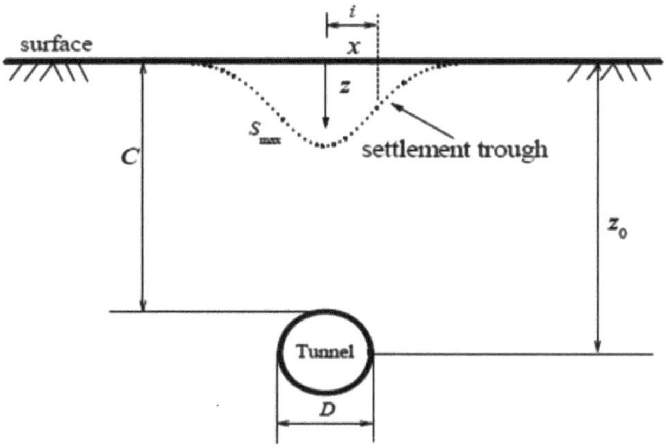

Fig. 2. Settlement trough, cited from Jin and Yang (2014)

To illustrate the effect of adjacent structure in modifying the settlement trough, two cases were analyzed using the PLAXIS model for tunnel construction with and without considering the building on top of the ground surface. It is shown in Fig. 3 that in case without building, the settlement trough is clearly obvious and it follows the settlement trough shown in Fig. 2. The second curve which is included in Fig. 3, represents the settlement trough in case of having building at ground surface. The shown value of settlement reflects the settlement during the tunnel construction (i.e. settlement due to building construction was reset to zero). It is shown from figure that the presence of building affected the common adòpted settlement trough shape. Settlement value in the zone under the building is clearly increased, and the building suffered not only from settlement but also from tilting of about 0.0025, which exceeds 1/500.

4 Results and Discussion

The first type of wall is unconnected piles, 0.5 m diameter. Piles spacing are $2D_{pile}$, $4D_{pile}$ and $6D_{pile}$, and piles length are $1D_{tunnel}$, $1.5D_{tunnel}$ to $2D_{tunnel}$. Model construction comprises 7 stages. Initial stage, building construction, wall construction,

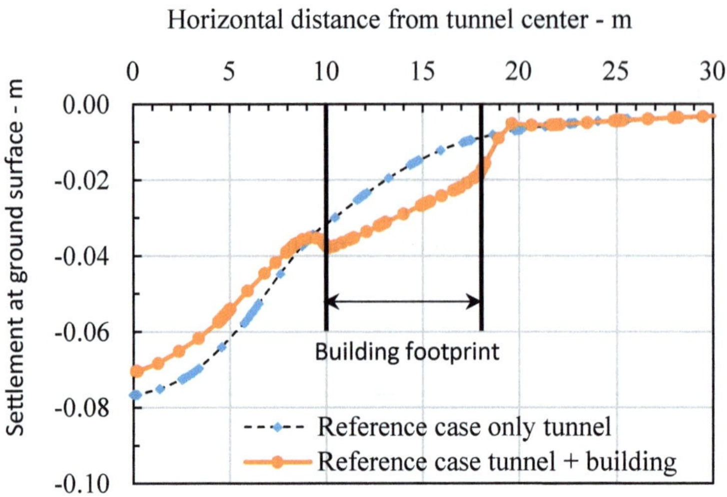

Fig. 3. Settlement trough at ground surface during the contraction stage

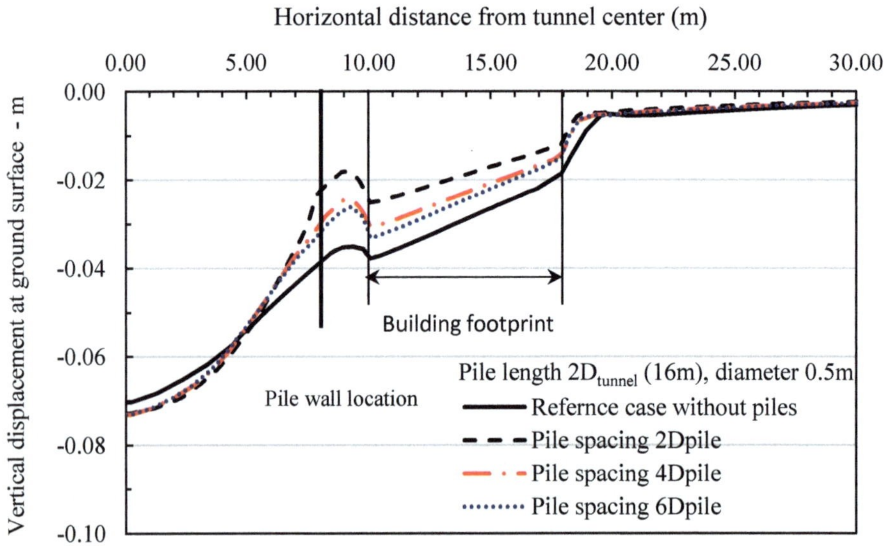

Fig. 4. Variation of vertical settlement at ground surface for the case of using pile wall length = $2D_{tunnel}$ (16 m), 0.5 m diameter in dry sand, for various pile spacing

tunnel excavation (TBM), volume loss due to tunneling, grouting and final lining. Settlement was reset to zero after building construction to investigate only the effect of tunneling construction. Settlement at ground surface after using the piles wall compared to reference case without wall are presented in Figs. 4, 5 and 6. It can be noticed

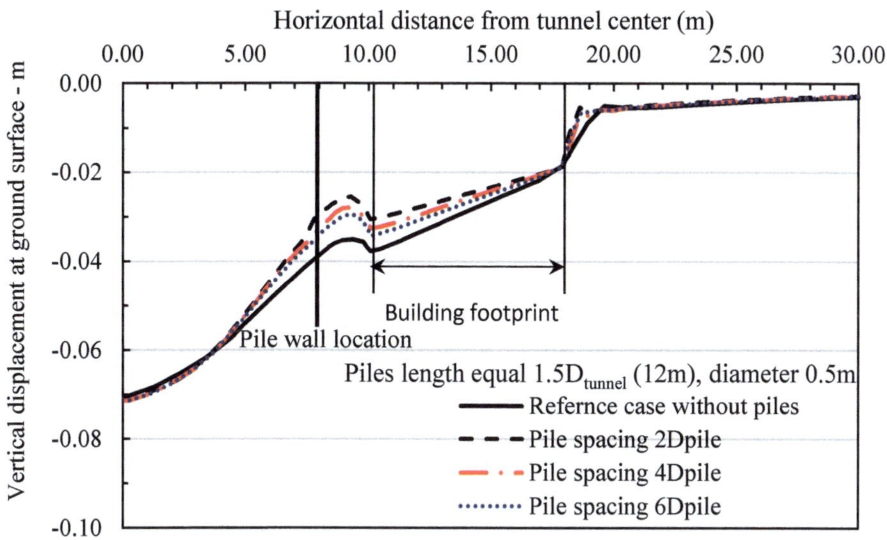

Fig. 5. Variation of vertical settlement at ground surface for the case of using pile wall length = $1.5D_{tunnel}$ (12 m), 0.5 m diameter in dry sand, for various pile spacing

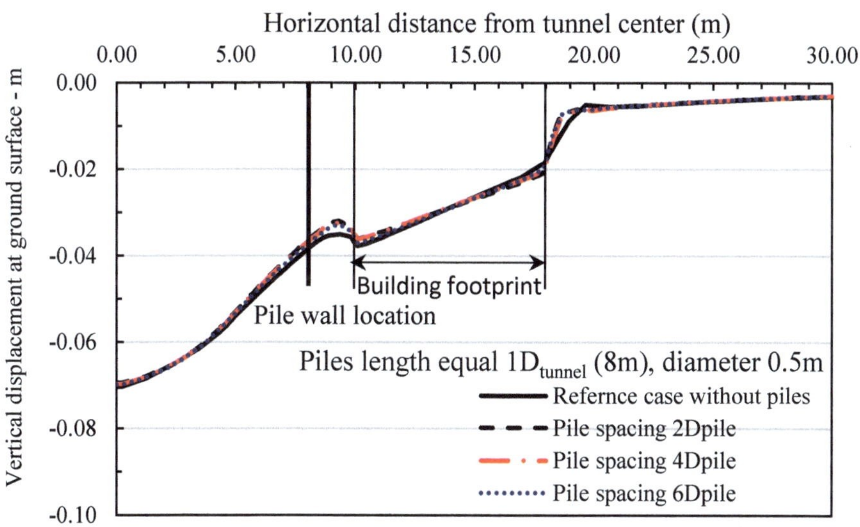

Fig. 6. Variation of vertical settlement at ground surface for the case of using pile wall length = $1D_{tunnel}$ (8 m), 0.5 m diameter in dry sand, for various pile spacing

from Fig. 4, for piles length equal $2D_{tunnel}$, that smaller spacing between piles provided more reduction in settlement. However, in Fig. 5, with shorter piles length of $1.5D_{tunnel}$, it can be noticed that the three values of piles spacing provided nearly the same value of settlement reduction. Yet, it is still better than the reference case (i.e. without piles wall). Finally, in Fig. 6 where piles length is only $1D_{tunnel}$, the wall didn't provide any noticeable reduction in settlement under the building.

For better evaluation of wall efficiency in reducing settlement under the building, an efficiency factor value (η) was calculated (after Bilotta and Russo 2011):

$$\eta = \left(\frac{S_r - S}{S_r}\right) \times 100 \tag{1}$$

where:

S: Vertical settlement in case of using pile wall or diaphragm wall at ground surface.

$S_{r:}$ Reference vertical settlement at the same location in case of not using wall.

The efficiency of piles wall in reducing the average settlement under the building for different cases of piles wall was calculated and presented in Fig. 7. It was found that 35% reduction in average settlement was achieved when using the piles spacing $2D_{pile}$, and piles length were extended to the bottom level of tunnel (i.e. $2D_{tunnel}$). Increasing the piles spacing to $4D_{pile}$ reduced the achieved reduction in settlement to about 21%. However, this reduction is higher than that achieved with pile spacing $2D_{pile}$, but piles length extended only to the tunnel center level or tunnel top level (i.e. $1.5D_{tunnel}$, $1D_{tunnel}$). The same trend was found when pile spacing increased to $6D_{pile}$. This means that increasing the piles length is more influential in settlement reduction more than the piles spacing.

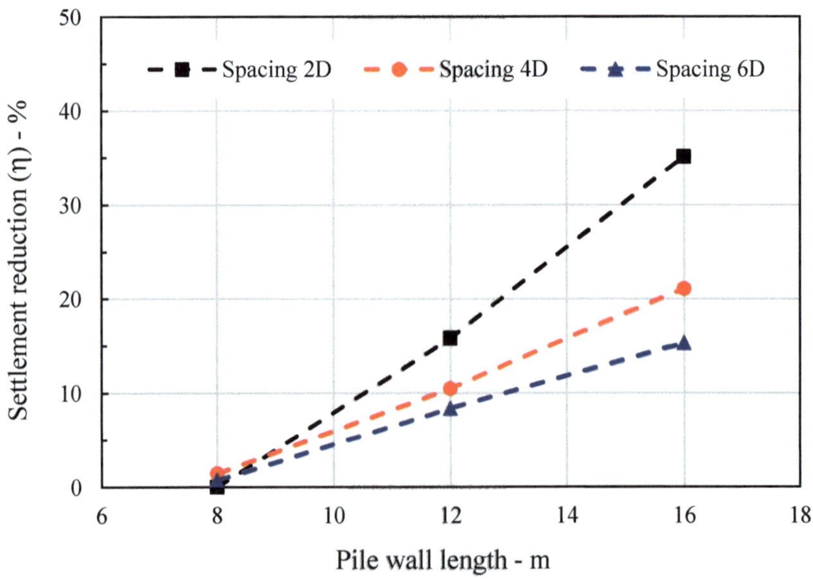

Fig. 7. Reduction in average settlement under the building (not connected piles wall, diameter 0.5 m)

Another group of piles wall were studied, that group has the same configurations except that it has a connection beam for the piles top. The connection beam's dimensions are (0.5 m width × 0.8 m height). Piles connection to the beam is considered as hinged connection and the beam was considered embedded in ground. Results of connected piles compared to not-connected piles are presented in Fig. 8. It is noticed that the connection between piles provided some more reduction in settlement for the different pile spacing. Yet, still the pile length has the most influential effect in decreasing the settlement.

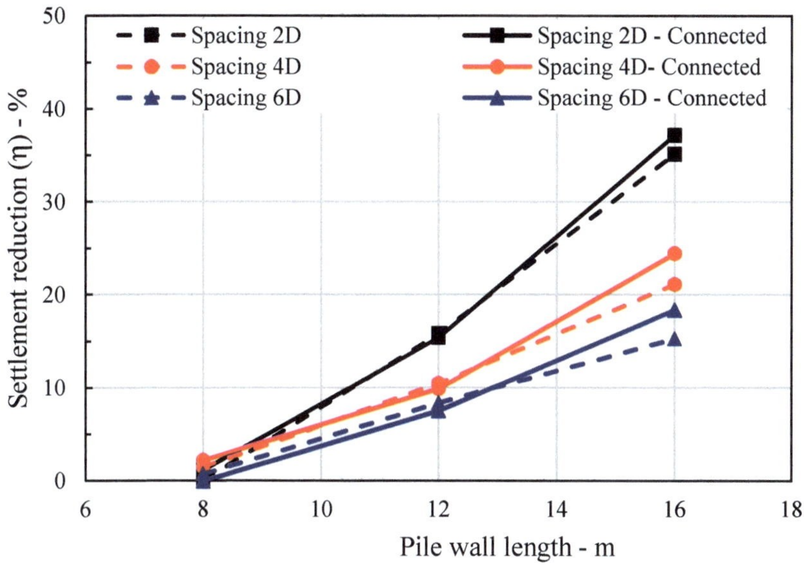

Fig. 8. Reduction in average settlement under the building (case of not connected piles and connected piles, piles diameter 0.5 m)

The effect of wall in reducing the building tilting was also investigated and the reduction in building tilting for different cases of piles spacing and length was calculated as a percentage from the reference tilting before using the piles wall. Results for connected and not-connected piles are presented in Fig. 9. It is shown that building tilting is reduced due to the use of piles wall. The reduction in tilting increased as the piles spacing decrease and the pile lengths increase. It is also noted that further reduction in building tilting is shown in the case of connected piles wall. It can be concluded that the connection beam at the top of the piles provided group action to the piles and reduced the horizontal displacement which subsequently reduced the building tilting.

Based on the above-mentioned, it can be concluded that extending the wall down to the level of tunnel bottom provided the best reduction in the average settlement, and connecting the piles contributed more in the reducing the horizontal displacement and consequently the reducing the building tilting. The piles connection altered the

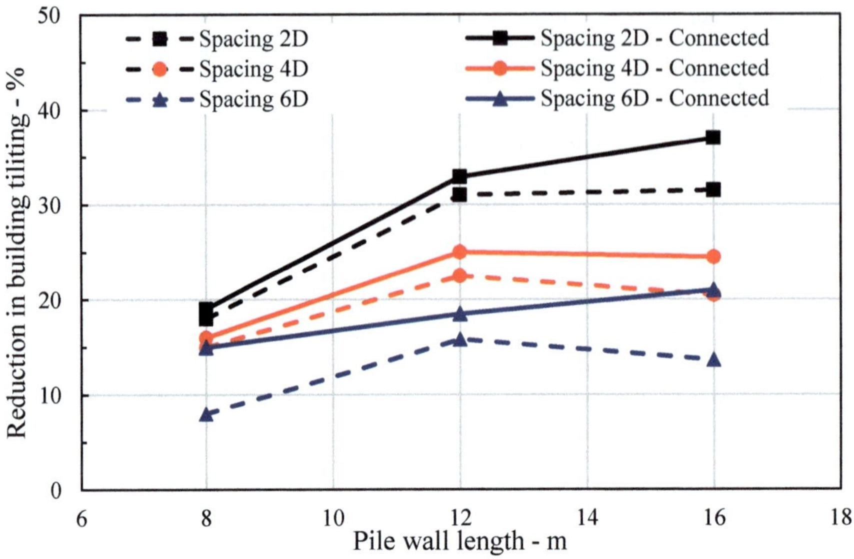

Fig. 9. Wall efficiency in reducing tilting under the building (case of not connected piles and connected piles, piles diameter 0.5 m)

behavior for the wide spacing piles wall as well. It is shown in Fig. 8 that for the 16 m wall, connected piles with $6D_{pile}$ spacing provided reduction in settlement nearly equal to the reduction in settlement in the case of not connected piles with $4D_{pile}$ spacing. From Fig. 9, it can be noticed that both, connected $6D_{pile}$ spacing wall and not-connected $4D_{pile}$ spacing wall, provided the same reduction in tilting. Considering that the differential settlement in most cases is more critical to the structure than the average settlement. Then, connecting the piles can be efficient in those cases even with wider spacing.

To investigate the effect of increasing the pile diameter, piles 0.6 m diameter was considered. In this group, only the case of piles wall length equal to $2D_{tunnel}$ was considered (i.e. 16 m), and it was analyzed for the different spacing $2D_{pile}$, $4D_{pile}$ and $6D_{pile}$ and for the two cases of piles connection. The connection beam in this case was (0.6 m width × 0.8 m depth) and the connection between piles and beam was hinged. Figure 10 presents a comparison between 50 and 60 cm diameter piles in reducing the building tilting. The figure also considers piles connection types. It is shown that for the smaller piles spacing (i.e. $2D_{pile}$) the increase in piles diameter improves the wall efficiency in reducing the building tilting and the connection between piles added to that improvement. However, for larger piles spacing (i.e. $6D_{pile}$), the governor in improving the wall efficiency is the pile connection rather than the pile diameter.

The second type of wall, considered in this study, is the diaphragm concrete wall. Efficiency of the wall in reducing the building tilting was calculated for walls with different heights and thicknesses, results are presented in Fig. 11. Results indicate that using continuous wall provided more reduction in the building tilting, as the wall terminated the soil particles movement and reduced the differential settlement effect

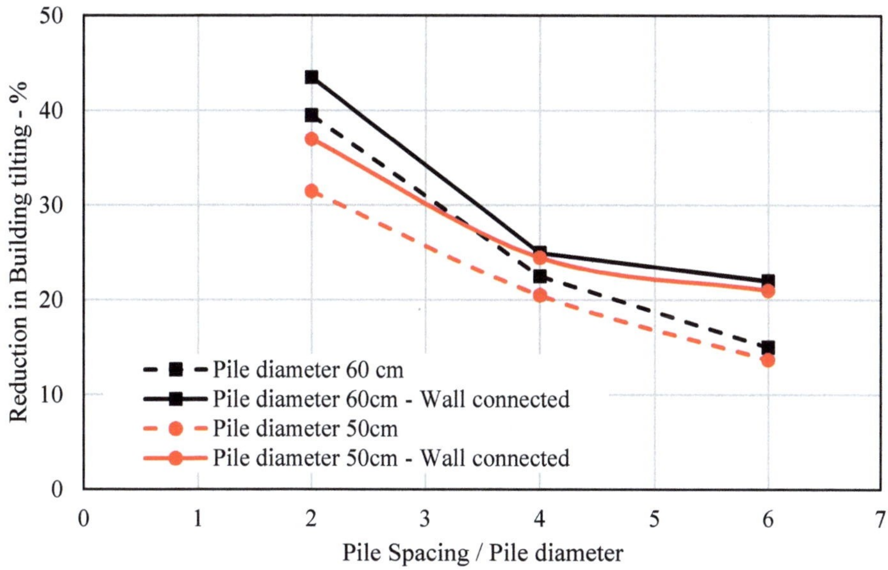

Fig. 10. Wall efficiency in reducing building tilting – pile length equal 2Dtunnel (16 m), (case of not connected piles and connected piles)

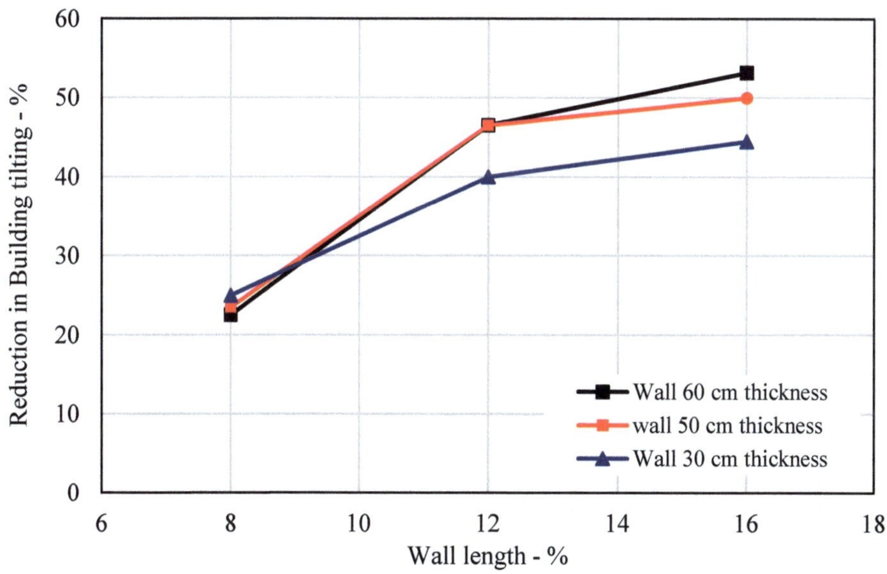

Fig. 11. Reduction in building tilting in case of using diaphragm wall

that could have reached the building. Increasing the wall length provided more reduction. It is also noticed from figure that the effect of wall thickness on reducing the building tilting starts to be more obvious in longer walls, $1.5D_{tunnel}$ and $2D_{tunnel}$ (i.e. 12 and 16 m).

To be able to demonstrate the best wall criteria in reducing the building tilting, the obtained reduction in building tilting is presented in the following Figs. 12 and 13. In Fig. 12, piles wall of spacing $4D_{pile}$ and $6D_{pile}$ with piles length 16 m are presented. It

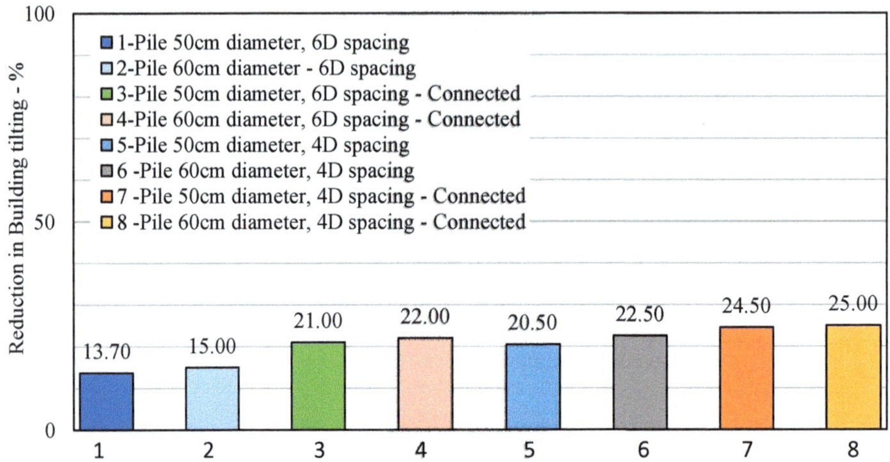

Fig. 12. Comparison between different wall configurations in reducing building tilting (piles length equal 2Dtunnel (16 m))

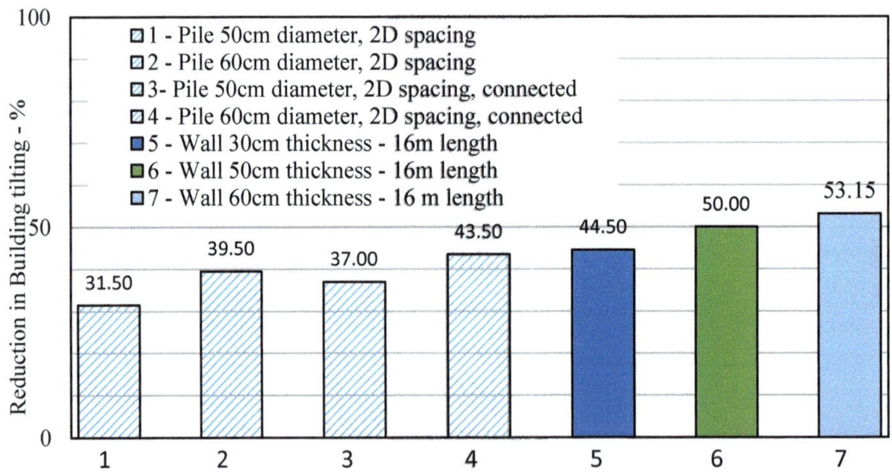

Fig. 13. Comparison between wall efficiency in reducing the building tilting (piles length equal 2Dtunnel (16 m))

is shown that for these piles spacing, and despite of the length of piles, the gained reduction in building tilting doesn't exceed 25% which is a relatively small value. Accordingly, those values of spacing are not recommended even with connection beam.

In Fig. 13, the reduction of building tilting is presented for the piles wall with spacing $2D_{pile}$ and length equal to $2D_{tunnel}$ (i.e. 16 m), against the diaphragm wall with different thicknesses having the same depth. It can be noticed that in case of using piles wall, with sufficient depth, the connection between piles allow using piles having smaller diameters (e.g. 50 cm piles with connection provided slightly different result than 60 cm piles which are not connected). The use of diaphragm wall with small thickness 30 cm didn't provide much reduction in tilting than the use of piles wall 60 cm diameter with connected piles. Nevertheless, increasing the wall thickness to 50 cm and 60 m provided more reduction in tilting.

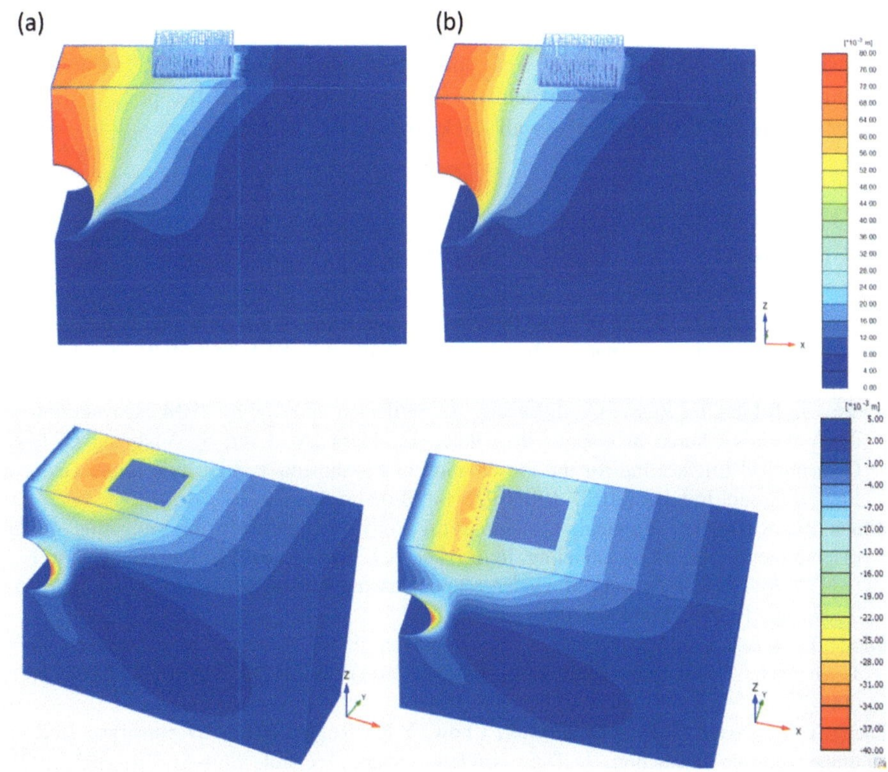

Fig. 14. (a) Total displacements contours due to tunnel construction for the case of (i) no wall installed and (ii) not-connected piles wall installed. (b) Horizontal displacement contours due to tunnel construction for the case of (i) no wall installed and (ii) not-connected piles wall installed

5 Conclusions

This study aims to provide a comparison between the different types of barrier walls (i.e. piles and diaphragm walls) that could be used to mitigate the effect of constructing a tunnel close to an existing building. The resulted settlements under the building were compared for different piles/diaphragm walls configurations and the following points were concluded.

1. Installing piles wall or diaphragm wall between the building and tunnel affected the soil movement in the zone above tunnel and contained most of the displacement in front of the wall, Fig. 14.
2. Increasing the piles length has a greater influence on the behavior and reduces the building tilt and average settlement more than the influence of reducing piles spacing or increasing pile diameter.
3. Using a connected piles wall having piles spacing of $2D_{pile}$ and depth of wall extended to the tunnel bottom provided the best configuration in case of using piles wall.
4. Using diaphragm wall, even with small thickness, provides good reduction in building tilting. However, to achieve higher efficiency than the piles wall, it is recommended to increase the wall thickness.

References

Alonso, E.E., Ledesma, A.: High speed railway tunnel crossing of downtown Barcelona: design, construction and performance. In: Proceeding of the XVI ECSMGE, Geotechnical Engineering for Infrastructure and Development (2015)

Aversa, S., Bilotta, E., Russo, G., di Luccio, A.: Settlement induced by TBM excavation under the Vittoria's Church in Napoli. In: Bilotta, E., Flora, A., Lirer, S., Viggiani, C. (eds.) Geotechnical Engineering for the Preservation of Monuments and Historic Sites. Taylor & Francis Group, London (2013). ISBN 978-1-138-00055-1

Bilotta, E., Russo, G., Viggiani, C.: Numerical study of a measure for mitigating ground displacements induced by tunneling. In: Bakker, K.J., Bezuijen, A., Broere, W., Kwast, E.A. (eds.) Geotechnical Aspects of Underground Construction in Soft Ground, pp. 357–362. Taylor and Francis, London (2006)

Bilotta, E., Russo, G.: Use of a line of piles to prevent damages induced by tunnel excavation. J. Geotech. Geoenvironmental Eng. **137**(3) (2011). doi:10.1061/(ASCE)GT.1943-5606.0000426

Cheng, C.Y., Dasari, G.R., Leung, C.F., Chow, Y.K., Rosser, H.B.: 3D numerical study of tunnel-soil-pile interaction. J. Tunn. Undergr. Space Technol. **19**(4–5), 381–382 (2004). doi:10.1016/j.tust.2004.02.011

Di Mariano, A., Gesto, J.M., Gens, A., Schwarz, H.: Ground deformation and mitigating measures associated with the excavation of a new Metro line. In: 14th ECSMGE. Millpress Science Publishers, Rotterdam, The Netherlad, vol. 4, pp. 1904–1906 (2007)

Dias, T., Bezuijen, A.: Pile tunnel interaction: literature review and data analysis. In: Proceeding of the World Tunnel Congress 2014, pp. 1–10 (2015)

El-Nahas, F.M., Ahmed, A.A., Esmail, K.A.: Prediction of ground subsidence above tunnels in Cairo. In: Proceedings of the Fourteenth International Conference on Soil Mechanics and Foundation Engineering, Hamburg, vol. 3, pp. 1453–1456 (1997)

El-Sayed, S.M.: Elasto-plastic three-dimensional analysis of shielded tunneling with special application on greater Cairo Metro. Ph.D. thesis, Ain Shams University, Faculty of Engineering, Cairo, Egypt (2001)

Gan, P., Tang, X., Zhao, Y., Wang, H.: Numerical analysis of effect of tunnel excavation on underground pipelines. In: International Conference on Pipelines and Trenchless Technology. ASCE (2014)

Gerheim, T., Dias, S., Bezuijen, A.: Data analysis of pile tunnel interaction. J. Geotech. Geoenvironmental Eng. (2015). doi:10.1061/(ASCE)GT.1943-5606.0001350. ASCE

Hu, X.Y., He, C., Jiang, Y.C., Wang, J., Zheng, H.L.: Research on the impact of shield tunneling on adjacent pile foundation using FEM. In: International Conference on Pipelines and Trenchless Technology. ASCE (2013). doi:10.1061/9780784413142.067

Jin, J., Yang, M.: Ground subsidence caused by tunnel excavation in sand. Tunn. Undergr. Constr., 174–183 (2014). doi:10.1061/9780784413449.018

Minh, N.V., Wout, B., Johan, B.: Volume loss in shallow tunneling. Tunn. Undergr. Space Technol. (2016). doi:10.1016/j.tust.2016.06.011

Peck, R.B.: Deep excavations and tunnelling in soft ground. In: Proceedings of the 7th International Conference on Soil Mechanics and Foundation Engineering, Mexico City, pp. 225–290 (1969)

PLAXIS 3D 2016. Plaxis bv, Delft, Netherlands (2016)

Schanz, T.: Zur Modellierung des Mechanichen Verhaltens von Reibungsmaterialen. Stuttgart Universitat, Habilitation (1998)

Gang, W.: Selection and distribution of ground loss ratio induced by shield tunnel construction. Chin. J. Geotech. Eng. J. **32**(9), 1354–1361 (2010)

The Monitoring of Segments Dislocation Deformation in Shield Tunnel Based on BOFDA

Shi Bin(✉) and Wang Xing

School of Earth Sciences and Engineering,
Nanjing University, Nanjing, Jiangsu, China
shibin@nju.edu.cn

Abstract. Shield tunnel is widely used in underground traffic engineering and underground pipeline structure, and the structure health monitoring is an important project to ensure the long-term health and safety of tunnels. Regarding to the deformation characteristics of shield tunnel, this paper presents a new method for the layout of sensing fiber based on BOFDA. By analyzing the plane geometry, the relationship between strain and deformation of segment dislocation was determined. The test accuracy and sensitivity of this method depends on fixed point distance. Laboratory test was conducted and the calculated results with this method had good consistency with the results by conventional dial meters. Using this long distance, high-density, high-accuracy and lost-cost health monitoring method, the safety of shield tunnel can be greatly ensured.

Keywords: Shield tunnel · Segment dislocation · BOFDA · Optical fiber sensing technology · Health monitoring

1 Introduction

Shield construction method is widely used in urban subway transportation construction due to its fast excavation, automatic operation, less effect on the ground traffic and other advantages. The structure control and health monitoring of the shield tunnel is of great significance to, on the one hand, the construction of the tunnel, and on the other hand, there are the increasing impacting factors of the surrounding construction to the tunnel [1, 2]. The shield segments are just like a "cell" of the tunnel and the deformation and stress of each "cell" has a close relationship with the overall deformation of the tunnel. Then, the deformation of the whole tunnel could be evaluated by the each segments deformation. The conventional geotechnical monitoring techniques are generally based on mechanical, hydraulic or electrical sensing elements and there are some inherent limitations. Most sensors for measuring deformation, temperature, and vibration are point (discrete) based sensors. The readings taken in the field are usually sensitive to electromagnetic interference. The quality of waterproof measures also significantly affects the durability (long-term performance) of these sensors [3, 4]. In order to capture tunnel deformations and evaluate stability conditions in real time, it is of great importance to develop an improved smart monitoring system which can overcome the above mentioned shortcomings.

S. Agaiby and P. Grasso (eds.), *Engineering Challenges for Sustainable Underground Use*, Sustainable Civil Infrastructures, DOI 10.1007/978-3-319-61636-0_16

Distributed fiber optic sensing technologies have emerged in the field of engineering monitoring since 1990s. This type of technology enables long-distance and distributed monitoring due to the flexibility of the sensing cables and they can be easily implanted into the surface or internal structure of concrete. The implantation methods of the sensing cables are simple and hence barely disrupts the overall construction process. There have been increasingly prevalent application of fiber optic sensing in the infrastructure control and health monitoring on the macro level [5–7], few studies, however, reported applications of such technology dealing with the three dimensional deformation of tunnel.

This study adopted a distributed fiber optic sensing technology for monitoring the concrete segments deformation of Suzhou metro tunnel, China. The effects of segment strips on the structural stress and the deformation are released. The study results facilitated verifying the feasibility and effectiveness of distributed fiber optic sensing technologies in structural health monitoring of a tunnel construction.

2 Principle of BOFDA

Figure 1 summarizes the principle of BOFDA: When a pump (pulse) laser and a probe laser is injected simultaneously into the two ends of an optical fiber, different scattering phenomena occur. One of the special scattering beams is termed as the stimulated Brillouin backscattering light, and its center frequency (CF) shifts linearly with axial compression (stretching) or temperature change of the optical fiber. This relationship can be expressed as:

$$\upsilon B(\varepsilon, T) = \upsilon B(\varepsilon 0, T0) + C1(\varepsilon - \varepsilon 0) + C2(T - T0) \tag{1}$$

where υB ($\varepsilon 0$, T0) denotes the initial CF; ε and $\varepsilon 0$ denote the strain change in the optical fiber and the initial strain, respectively; T0 and T denote temperature before and after straining or thermal change, respectively; C1 denotes the strain coefficient, which is 0.05 MHz/$\mu\varepsilon$; and C2 denotes the temperature coefficient, and is 1.2 MHz/°C. By counting the traveling time of scattering light in the fiber, locations of the stain or temperature events can be determined.

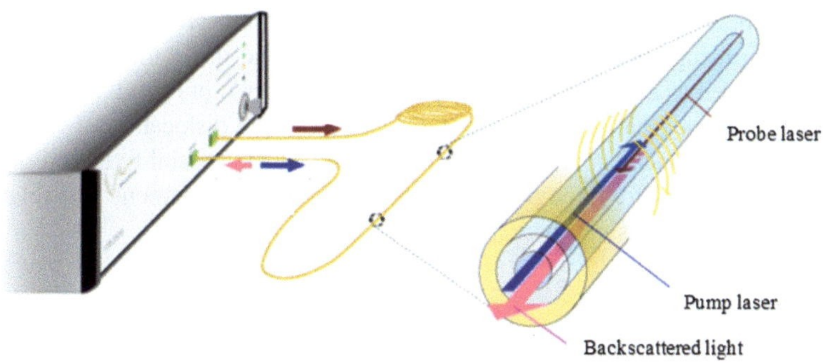

Fig. 1. The principle of BOFDA (adopted from fibrisTerre fTB series manual)

The difference between BOFDA and other optical fiber sensing technology is that a baseband transmission function (BTF) can be demodulated from BOFDA, and considerably higher demodulation precision and spatial resolutions can be determined following the deconvolution procedure with the BTF. An fTB2505 demodulator manufactured by fibrisTerre can achieve ±2 $\mu\varepsilon$ strain accuracy and 20 cm spatial resolution, which satisfies the requirements for shield tunnel monitoring.

For field applications, cares should be taken to protect the delicate sensing cables: the sensing cable should be well protected with a specially designed jacket. To monitor the deformation of a structure, the cables were installed on the structure surface with a matched fixture to ensure that the fiber deformed in a compatible manner with the structure. The demodulation of strain change in the fiber helps in obtaining the health information of the entire structure. Currently, the maximum test distance of BOFDA is 50 km [8, 9].

3 The Sensing Mode of Shield Tunnel Deformation

3.1 Deformation Mode Analysis

Fibers are only sensitive to the deformation occurring along the axial direction, and the deformation direction of the segment of the shield tunnel is orthogonal to the fiber arranged in the axial direction of the tunnel. The deformation of the shield tunnel cannot cause the tensile and compressive deformation of the fiber. And, it is necessary to establish the relationship between the amount of optical fiber deformation and the amount of block to achieve, if you consider the use of optical fiber distribution and high test accuracy to achieve the advantages of the entire shield tunnel section of the deformation.

To build up this relationship, at first, the optical fiber demodulation devices was carried out to transform optical fiber strain into the strain in the fiber. The relationship between the amount of deformation and the strain can be expressed by Eq. (2):

$$L = \varepsilon \cdot l \tag{2}$$

where L is the deformation, ε for the test to obtain fiber strain, l for the occurrence of two-point distance between the strain. After fixing the fix-ended cables, the relative deformation between the two fixed points can be measured [16].

The relationship between the deformation of fiber and the deformation of segment can be considered as a planar geometric problem. As shown in Fig. 2, the sensing fiber is arranged in a way that is oblique to the direction of the dislocation of the tube, one end fixed on the shield segment, the other end fixed on the special fixed rod, a, b, c for the fiber fixed point, d for fixed position of the bar and shield segment. The relationship between the fixed points can be expressed as:

$$L_3 = \sqrt{L_2^2 - L_1^2} \tag{3}$$

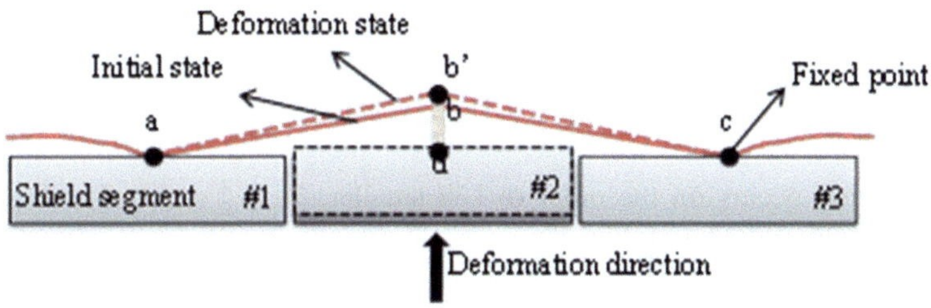

Fig. 2. Schematic diagram of the deformation of the adjacent segments

where L_1 is the distance between the fixed points a and d; L_2 is the distance between the fixed points a and b; and L_3 is the distance between b and d.

When the #2 segment is deformed against the #1 and #3 segments, the fixed rod drives the fiber from the initial position b to b', the distance relationship between the fixed points can be expressed by (4):

$$L_3' = \sqrt{L_2'^2 - L_1^2} \tag{4}$$

where L_2' is the distance between a and b′; L_3' is the distance between d and b′. According to Eqs. (3) and (4), the deformation amount δ of the segment can be calculated by the Eq. (5):

$$\delta = L_3' - L_3 \tag{5}$$

Combined (2) with (5), we can get the relationship between the deformation of the shield segment and the measured fiber strain.

$$\delta = \sqrt{(1+\varepsilon)^2 L_2^2 - L_1^2} - \sqrt{L_2^2 - L_1^2} \tag{6}$$

where δ is the amount of deformation, ε is the axial strain value of the fiber, L_1 and L_2 are the horizontal distance and the straight line distance between the fixed points of the two fibers, respectively, and the initial fixed position of the fiber is fixed value, which can be obtained between the fiber strain and the deformation of the shield tunnel.

3.2 Analysis of Error, Sensitivity and Test Range

The formed subway tunnel acceptance specification: the control range of the adjacent segment radial control table is generally 0–10 mm, adjacent shield segment control range is generally 0–15 mm, the test accuracy required is between ±0.01 mm [15]. The Eq. (6) is reduced to:

$$\delta = \frac{2\varepsilon L_3^2}{\sqrt{2\varepsilon L_3^2 + L_2^2} + L_2} \tag{7}$$

As the strain occurs on the order of l, h smaller, $2\varepsilon \cdot L_2^2 \ll L_3^2$, The Eq. (7) can continue to be simplified as follows:

$$\delta = \frac{L_3^2}{L_2} \cdot \varepsilon \tag{8}$$

It can be seen from (8) that the error of deformation and the magnitude of strain ε are proportional to L_2^2, and the vertical distance L_3 between two fixed points is inversely proportional. L_2 and L_3 are fixed values, so the error of the deformation of the shield segment is only related to the strain test accuracy. In this case, the ratio of L_3^2 to L_2 is defined as the sensitivity coefficient k, and the higher the sensitivity coefficient, the higher the sensitivity of the whole system. The strain in this calculation method is the strain average of the test fiber:

$$\varepsilon = \frac{\varepsilon_1 + \varepsilon_2 + \ldots + \varepsilon_n}{n}, \left(n = \frac{L}{h}\right) \tag{9}$$

In the equation, L is the fixed pitch of the fiber, and h is the sampling interval of the test instrument. To eliminate the spatial resolution SR (Spatial Resolution), we take the middle of the test data L and the length of the SR to determine the strain mean. Reducing the system error requires a corresponding increase in strain demodulation accuracy.

The system sensitivity of the method is determined by the size of L_2 and L_3, and its test sensitivity can be improved by increasing the value of L_3^2/L_2. In practical application where considering the tunnel clearance requirements, L_2 cannot be too large (generally <0.5 m; Considering the shield tunnel segment width, L_3 distance cannot be too small, usually 1.2 m.

The test range of the whole system is determined by the maximum tensile strain (±15000 με) and the fixed length of the fiber. Take the L_2 value as 120 cm and the L_3 value as 20 cm, the system error, sensitivity and scale of the method under different performance parameters of the distributed demodulator are shown in Table 1. It can be

Table 1. Different performance parameters under the test equipment system error, sensitivity and test range.

Technical index	BOFDA	BOTDA	BOTDR
	fTB2505	NBX-6050A	N8511
Strain accuracy (με)	±2	±10	±50
Error range (mm)	±0.01	±0.03	±0.15
Sensitivity (mm/ε)	0.03	0.03	0.03
Measurement range (cm)	−1–1	−1–1	−1–1

seen from Table 1, demodulation instrument demodulation accuracy is higher, the smaller the test error; The sensitivity and range of the test can meet the requirements of the site [10].

4 Test Verification

4.1 Test Plan

Based on the above analysis, it can be drawn that different fixed points have different effects on the test sensitivity. The calibration of sensitivity coefficient k should be conducted before the practical application. To verify the feasibility of this method and determine the coefficient k, the simulation test of segment distortion was carried out in laboratory. As shown in Fig. 3, two tough boards was fixed on two three-dimension movement equipment respectively, which can simulate two segments distortion movement. For setting different fixed heights, another punching steel plate was fixed on the moving board. Moving steel plate up and down the movement will lead to different displacement in sensing fiber. NZS-DSS-C06 strain-sensing fiber is used for the simulation test. A layer of polyurethane sheath is added to the outside of the cladding. And the outermost layer is 0.9 mm in diameter and its profile is shown in Fig. 4. The physical and mechanical property of the fiber is shown in Table 2. The fiber optic strain analyzer fTB2505 was by fibrisTerre in the German, the parameters of the unit are shown in Table 3.

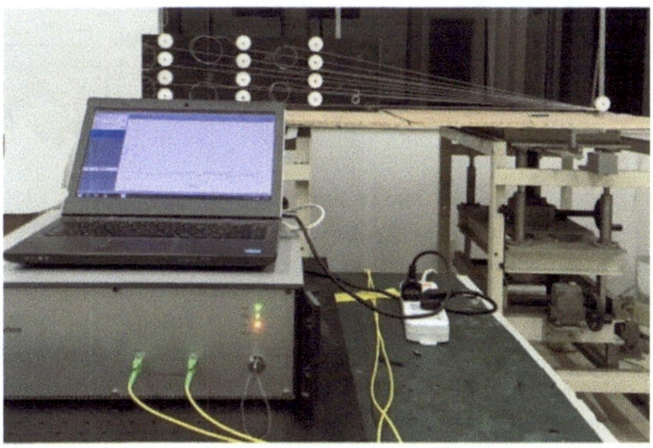

Fig. 3. Laboratory test set up

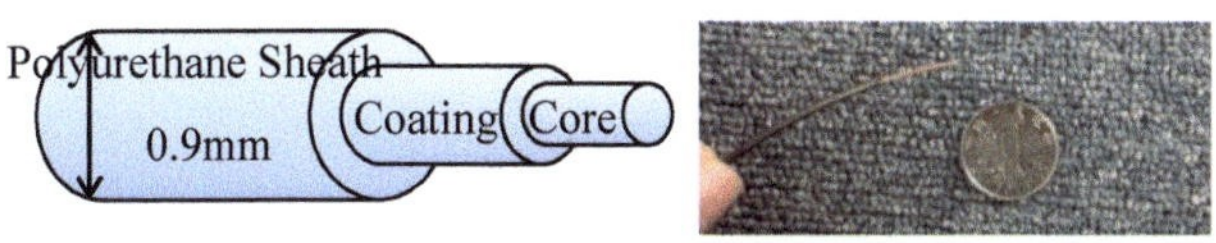

Fig. 4. The sensing fiber

Table 2. The parameter of sensing fiber

Name	No.	Core type	Diameter	Quality	Wavelength	Refractive index
Tight-buffed strain sensing fiber	NZS-DSS-C06	SM G.625b	0.9 mm	1.5 kg/km	1.3–1.5 μm	1.468

Table 3. The parameter of fTB2505

Name	Strain accuracy	Spatial resolution	Sampling resolution	Test distance	Weight	Max. consumption
fTB2505	±2 με	20 cm	5 cm	25 km	13 kg	60 W

The layout scheme of the different fixed-points sensing fiber is shown in Fig. 5. L_2 should longer than the segment length 120 cm, in this simulation, L_2 is 120 cm, 150 cm, 180 cm, respectively. And L_3 is 10 cm, 20 cm, 30 cm, 40 cm, respectively. All the sensing fiber was connected together to the equipment connectors.

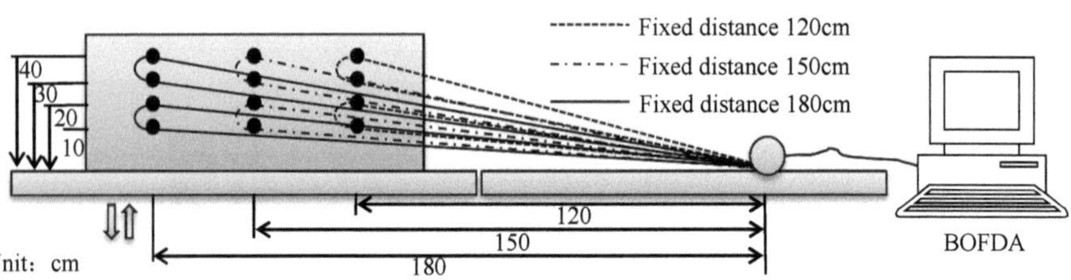

Fig. 5. The layout of sensing fiber

4.2 Procedure

The Sensing fiber need to be pre-pull before installation, the amount of pre-pull can be determined according to the actual situation. Figure 6 shows the initial strain of the sensing fiber, the amount of pre-pull is about ±200 με. From left to right, the distance of fixed points are 180 cm, 150 cm, 120 cm, respectively, and the lifting height are 10 cm, 20 cm, 30 cm, 40 cm. There are ten steps during the whole test, from 0–9 mm, each steps levels up 1 mm. During the lifting process, the fiber strain data is recorded every 15 min after the first level of stretching, in the meanwhile recording the displacement data of the dial meter.

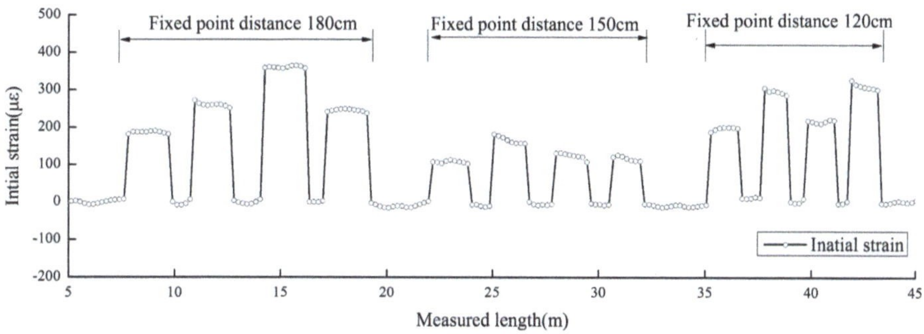

Fig. 6. The initial strain of sensing fiber

4.3 Test Result and Analysis

Figure 7 shows the strain changes during the test. The maximum strain values corresponding to the sections are: 1069 με, 870 με, 561 με, 297 με, 1508 με, 1170 με, 802 με, 416 με, 2272 με, 1789 με, 1240 με, 650 με. With the increase of the lifting capacity, the strain value of the sensing fiber is also increased step by step, but the larger amplitude increases with the point distance becomes smaller. For the same point distance, with the increase of the fixed point height, the fiber strain value increases in turn smaller. Figure 8 shows deformation curve of measured strain value in sensing fiber under the same point distance. It can be seen from the figure that with the increase of deformation, the strain coefficient of the measured fiber increases linearly, and the linear correlation coefficient R^2 is more than 0.99, which indicates that the linearity is good.

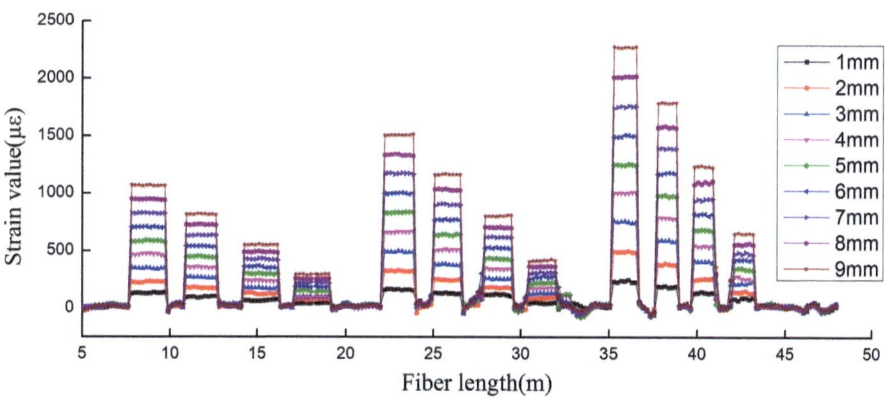

Fig. 7. Strain in sensing fiber under different displacement

The displacement can be calculated by fiber strain value using Eq. 8. Table 4 shows the deformation of the optical fiber strain value at the deformation level of the above points. It can be seen from the table that the deformation of the fiber strain test value is basically the same as the actual deformation, the maximum error of less than 10%, which can meet the requirements of shield tunnel monitoring.

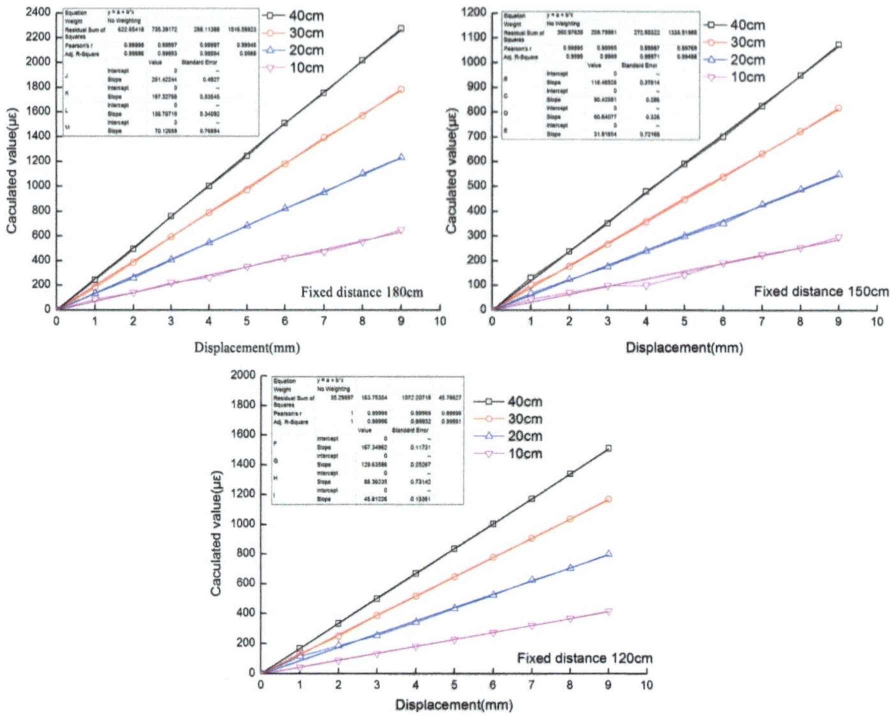

Fig. 8. The strain value in sensing fiber under same fixed point distance

Table 4. Calculated value under different condition

Disp./mm	120				150				180			
	10	20	30	40	10	20	30	40	10	20	30	40
1	1.01	1.12	0.98	1.08	1.09	1.01	0.97	1.10	1.07	1.03	1.03	1.06
2	2.02	2.10	2.08	2.06	2.02	2.00	2.09	1.95	2.10	2.01	2.00	2.06
3	3.08	3.09	3.00	3.01	2.95	3.01	3.00	3.04	3.02	3.05	3.10	3.03
4	4.01	4.06	4.11	4.16	4.00	3.90	4.01	4.06	4.09	4.01	4.01	4.03
5	5.10	5.05	5.06	5.05	5.00	5.05	5.12	4.90	5.00	5.01	5.05	5.09
6	5.90	6.01	6.06	6.09	6.00	6.10	6.01	6.06	6.10	5.98	6.01	6.03
7	7.06	7.02	7.02	7.00	7.10	6.98	7.03	7.01	7.00	7.06	7.06	7.00
8	8.00	8.12	8.10	8.10	8.06	8.01	8.09	7.99	8.00	8.01	8.01	8.09
9	9.01	9.06	9.10	9.12	9.09	8.90	9.02	9.02	9.10	9.12	9.06	9.07

5 Application Prospect

From the above theoretical analysis and test results, it can be concluded that the error of the test results of this BOFDA based segment distortion monitoring method is mainly effected by instrument system error and layout of the fiber. The main methods of reducing the error are:

(1) Improve the accuracy of fiber demodulation instrument.

The error of distributed optical fiber demodulator is caused by stability of laser source and the photosensitive element. It can be corrected by connecting with a constant temperature and strain fiber to offset the resulting shift. There are always strain transition sections in distributed optical fiber sensing technology, and the effecting area and error will get much bigger with the strain value getting smaller. This effect can be reduced by improving the strain value.

(2) Layout of the sensing fiber.

The test accuracy will also affected by the quality of epoxy resin and paste during the layout process of sensing fiber. To get the good results, the distributed strain value should be even along the layout section. The strength of fixture should be high enough to support the fiber displacement in case of extra error.

6 Conclusions

In this paper, the distributed optical fiber sensing technology based on BOFDA is applied to monitor the distortion of segments deformation. By designing a special fix method, the strain value in fiber can connected with the distortion displacement. The sensitivity of the method is related to the spacing of the optical fiber. The theoretical test precision and the test range can meet the requirements of tunnel deformation monitoring. The feasibility of the above method and the validity of the test result are verified by laboratory test. In the monitoring application of this technology, the specific scope of the tunnel and the scope of other pipeline layout requirements should be considered previously.

Acknowledgement. Financial support from Scientific Research Project of Suzhou Rail Transit Tunnel Health Monitoring (SZZG06YJ1021002) is greatly appreciated.

References

1. Schotte, K., De Backer, H., Van Bogaert, P.: Strain measurements in precast concrete segments of a shield-driven tunnel. In: FIB Symposium on Concrete Engineering for Excellence and Efficiency, Prague (2011)
2. Tang, E., Lui, V., Wong, A.: Application of automatic deformation monitoring system for Hong Kong railway monitoring. In: FIG Working Week (2007)
3. Bassett, R.H., Kimmance, J.P., Rasmussen, C.: An automated electrolevel deformation monitoring system for tunnels. Geotech. Eng. (1999). Practice of Shanghai metro monitoring. Undergr. Eng. Tunn.
4. Mohamad, H., Soga, K., Bennett, P.J., Mair, R.J., Lim, C.S.: Monitoring twin tunnel interaction using distributed optical fiber strain measurements. J. Geotech. Geoenvironmental Eng. (2012) doi:10.1061/(asce)gt.1943-5606.0000656

5. Moffat, R., Sotomayor, J., Beltrán, J.F.: Estimating tunnel wall displacements using a simple sensor based on a Brillouin optical time domain reflectometer apparatus. Int. J. Rock Mech. Min. Sci. (2015). doi:10.1016/j.ijrmms.2014.10.013
6. Shi, B., Xu, H., Chen, B., Zhang, D., Ding, Y., Cui, H., Gao, J.: A feasibility study on the application of fiber-optic distributed sensors for strain measurement in the Taiwan strait tunnel project. Mar. Georesour. Geotechnol. (2003)
7. Bernini, R., Minardo, A., Zeni, L.: Reconstruction technique for stimulated Brillouin scattering distributed fiber-optic sensors. Opt. Eng. (2002)
8. Nöther, N.: Distributed Fiber Sensors in River Embankments: Advancing and Implementing the Brillouin Optical Frequency Domain Analysis. BAM (2010)
9. Mao, J.H., Jin, W.L., He, Y., Cleland, D.J., Bai, Y.: A novel method of embedding distributed optical fiber sensors for structural health monitoring. Smart Mater. Struct. (2011). doi:10.1088/0964-1726/20/12/125018

Erratum to: Mitigating Foundation Settlement Induced by Tunnel Construction

Mona Badr El-Din Anwar(✉)

Civil Engineering Department, The German University in Cairo, New Cairo, Egypt
mona.anwar@guc.edu.eg

Erratum to: Chapter "Mitigating Foundation Settlement Induced by Tunnel Construction" in: S. Agaiby and P. Grasso (eds.), *Engineering Challenges for Sustainable Underground Use*, Sustainable Civil Infrastructures, DOI 10.1007/978-3-319-61636-0_15

In the original version of this book, the author's second affiliation has to be deleted in Chapter 15, which is a belated correction. The erratum chapter and the book have been updated with the change.

The updated online version of this chapter can be found at
http://dx.doi.org/10.1007/978-3-319-61636-0_15

S. Agaiby and P. Grasso (eds.), *Engineering Challenges for Sustainable Underground Use*,
Sustainable Civil Infrastructures, DOI 10.1007/978-3-319-61636-0_17

Author Index

S. Agaiby and P. Grasso (eds.), *Engineering Challenges for Sustainable Underground Use*,
Sustainable Civil Infrastructures, DOI 10.1007/978-3-319-61636-0

MIX
Papier aus verantwortungsvollen Quellen
Paper from responsible sources
FSC® C105338

If you have any concerns about our products,
you can contact us on
ProductSafety@springernature.com

In case Publisher is established outside the EU,
the EU authorized representative is:
Springer Nature Customer Service Center GmbH
Europaplatz 3, 69115 Heidelberg, Germany

Printed by Libri Plureos GmbH
in Hamburg, Germany